T0255324

THE SOCIAL EVOLUTION OF HUMAN NATURE

This book sheds new light on the question of how the human mind evolved. Harry Smit argues that current studies of this problem misguidedly try to solve it by using variants of the Cartesian conception of the mind, and shows that combining the Aristotelian conception with Darwin's theory provides us with far more interesting answers. He discusses the core problem of how we can understand language evolution in terms of inclusive fitness theory, and investigates how scientific and conceptual insights can be integrated into one explanatory framework, which he contrasts with the alternative Cartesian-derived framework. He then explores the differences between these explanatory frameworks with reference to cooperation and conflict at different levels of biological organization, the evolution of communicative behaviour, the human mind, language and moral behaviour. His book will interest advanced students and scholars in a range of subjects including philosophy, biology and psychology.

HARRY SMIT is Assistant Professor in the Department of Cognitive Neuroscience at Maastricht University.

THE SOCIAL EVOLUTION
OF HUMAN NATURE

From biology to language

HARRY SMIT

Department of Cognitive Neuroscience, Faculty of Psychology and Neuroscience, Maastricht University

CAMBRIDGE
UNIVERSITY PRESS

CAMBRIDGE
UNIVERSITY PRESS

University Printing House, Cambridge CB2 8BS, United Kingdom

One Liberty Plaza, 20th Floor, New York, NY 10006, USA

477 Williamstown Road, Port Melbourne, VIC 3207, Australia

4843/24, 2nd Floor, Ansari Road, Daryaganj, Delhi - 110002, India

79 Anson Road, #06-04/06, Singapore 079906

Cambridge University Press is part of the University of Cambridge.

It furthers the University's mission by disseminating knowledge in the pursuit of
education, learning and research at the highest international levels of excellence.

www.cambridge.org
Information on this title: www.cambridge.org/9781107697553

© Harry Smit 2014

First published 2014
First paperback edition 2017

A catalogue record for this publication is available from the British Library

Library of Congress Cataloging in Publication data
Smit, Harry.
The social evolution of human nature : from biology to language / Harry Smit.
pages   cm
ISBN 978-1-107-05519-3 (Hardback)
1. Social evolution.   2. Cognition–Social aspects.   3. Social psychology.   I. Title.
HM626.S62 2014
302–dc23
2013036077

ISBN 978-1-107-05519-3 Hardback
ISBN 978-1-107-69755-3 Paperback

For Judith, Friso and Simon

... it is a heuristic maxim that the truth lies not in one of the two disputed views but in some third possibility which has not yet been thought of, which we can only discover by rejecting something assumed as obvious by both the disputants ...

F. P. Ramsey, *Foundations: essays in philosophy, logic, mathematics and economics*, edited by D. H. Mellor. London: Routledge & Kegan Paul, 1978, 20–21.

Contents

Preface and acknowledgments

This book combines scientific and conceptual insights. There is a simple reason: its author believes that both are indispensable for understanding human nature. I make here some autobiographical remarks in order to clarify how I arrived at this position.

I am trained as a biologist and was and still am interested in zoology and genetics. The biological approach to the behaviour of animals including humans had my interest, and at the University of Groningen I studied foraging behaviour of birds (starlings), social behaviour of monkeys (hamadryas baboons) and the principles of evolutionary genetics. Like many others, I was impressed by Hamilton's contribution to our understanding of the evolutionary genetics of social behaviour. But I found it hard to apply Hamilton's insights to human social behaviour because it was unclear to me how genetics can explain unique features of the human species like self-consciousness and language. I thought at that time that, in order to understand the origin of human nature, we first have to understand the evolution of these unique features of the human mind. But I had no idea of how to investigate these features and only noted that the mind/body problem somehow obstructed the application of biological principles to humans. The longer I thought about these problems and the more I read, the more confused I got. I decided to take some courses in philosophy in order to understand problems better. The result was that I became a philosopher as well, understood some problems better, but still did not understand much of the evolution of human nature.

After I graduated I had the opportunity to do a project on evolutionary epistemology at the University of Nijmegen. I studied evolutionary theory, possible applications to the evolution of cognition (and theories) and some parts of Kant's philosophy, because Konrad Lorenz made in his *Behind the mirror* a link between Kant's philosophy and evolutionary epistemology. It was in this period, presumably spring 1984, that a colleague invited me to join him to attend a lecture of the philosopher Peter Hacker, because he knew that I had an interest in the philosophy of mind. I remember clearly that Hacker explained, convincingly to me, why some cognitive psychologists and neuroscientists have it wrong when they argue that there are symbolic representations somewhere and somehow encoded in brain processes. I also remember that my colleague thought otherwise and raised his hand to make some objections. But there was something else that caught my attention, namely that Hacker made a simple observation: we can use a ripe tomato (as a sample) for explaining the word 'red', but there is no such thing as explaining a mental predicate (e.g. 'pain') through pointing at a sample in the brain or mind. Why was I then struck by this simple truism? With the benefit of hindsight, I can see two reasons.

First, because I assumed at that time that the idea of the logical empiricists that words are connected to reality through ostensive definitions was mistaken. Yet Hacker argued that ostensive definitions are useful for explaining the meaning of some words (although there are others ways of explaining word meaning) and that many philosophers misunderstood their role. There was a link here with evolutionary epistemology, for it was assumed in this field that we have innate knowledge of concepts enabling us to understand the use of words. Kant's philosophy was thought to be of potential interest, for Kant argued that the mind imposes regularities upon the external world (through concepts guiding our experiences). Extending Kant's ideas, Lorenz suggested that (innate) concepts of our mind are adapted to the environment, just as the eye is adapted to the rays of the sun. Whereas Hacker's ideas raised the problem of how we can understand the evolution of the use of words as an extension of the use of gestures such as pointing, Lorenz's ideas raised the problem of whether our knowledge of concepts evolved as the result of mutations affecting brain processes. I realized then that one cannot have it both ways.

The second reason was because Hacker emphasized that there is no such thing as explaining a psychological predicate (e.g. 'pain') through pointing at a mental sample (in the immaterial mind). He argued that the assumption that the physical and mental domains are comparable to each other is misguided. The mental domain is, in contrast to the physical domain, not populated with immaterial objects, events, states or processes which can be observed through introspection (enabling us to report what we discover in the mind). This observation raised the question of how children learn the use of psychological predicates if it does not rest on introspection. Children learn their use when they learn to extend and replace natural expressions of psychological phenomena (also displayed by our closest relatives) with linguistic ones (e.g. 'Au!', 'It hurts', 'I have a pain' instead of crying or screaming out of pain) *and* when they comprehend that these linguistic expressions uttered by others are manifestations of these phenomena (enabling them to say: 'He has a pain'). Linguistic expressions of psychological phenomena do not rest on introspection, while attributing psychological predicates to others rests on what they say and do. Again, I realized that Hacker's ideas had implications for how we should investigate the evolution of the way we communicate about mental phenomena.

I have to add that I was at that time only dimly aware that the arguments Hacker advanced had an enormous impact on how we should conceive of human nature. But I did realize that Hacker discussed an alternative view that could be far more interesting than the theories I had been studying. I went to the library the next day and obtained a copy of Hacker's *Insight and illusion*, which is about Kant's, Frege's, Russell's and Wittgenstein's early and late philosophy. After I finished reading (convincing me that there was an alternative view), I decided to visit Hacker in Oxford in order to discuss some problems. I can hardly recall what we discussed then, but remember one analogy that Hacker used for explaining why the rules for the use of words are autonomous. I found it so extremely illuminating that, after I returned home, I spent about two days explaining the analogy (and its implications) to my girlfriend (although she understood the essential point probably within a few minutes). Hacker remarked that the rules of tennis *would* be different if we had invented the game while living on the moon, but the rules of tennis (we have invented on earth) are not accounted for by referring to the laws of gravitation, for we *could* have invented different rules.

After finishing the project in 1985 and later my PhD in 1989 (which was about the nature/nurture problem in immunology, developmental biology and psychology), I decided to return to biology. I had read Leo Buss' *The evolution of individuality* and, because of my background in immunology and developmental biology, decided to learn more about evolutionary transitions, i.e. the transition from unicellular creatures to the symbiotic unicell, and from the symbiotic unicell to multicellular organisms. This was at the time an interesting research topic because there was for the first time conclusive evidence that mitochondria and chloroplasts (living inside host cells) were originally free-living bacteria and later became endosymbionts. Molecular biology was providing us with data enabling us to get definite answers to old questions. It was also clear that these data about subcellular structures and molecules were going to transform our ideas about the origin and evolution of the early forms of life. Buss' book was interesting because he noted that there was at the subcellular level far more potential for conflict than was realized by biologists at that time. Moreover, he argued that cooperation between cell organelles and host cells, and between cells in a multicellular organism, require explanations.

While studying cooperation and conflict at a subcellular level in 1991, one particular example caught my attention. My girlfriend, then a gynaecologist in training, told me that the rate of spontaneous abortions was much higher than expected (closer to 30% than to 5%). And she added that the embryo produced the hormone hCG (human Chorionic Gonadotropin; associated with pregnancy sickness) affecting female physiology: it suppresses the shedding of the uterine lining (endometrium) and hence prevents menstruation. When I read some papers I realized that the production of hCG by the embryo raised interesting evolutionary questions, for it showed that embryos determine in part their own chances of survival. But why do embryos determine their own fate? Why do they manipulate maternal physiology? These were new questions to me, because I had thought about maternal–foetal interactions in the context of immunology. Because the embryo is genetically not identical to the mother, immunologists studied the mechanisms that prevent an immune response of the mother (presupposing cooperation between mother and child). However, I was also familiar with the work of Robert Trivers on parent–offspring conflict, and I realized that there could be a conflict involved between parent and offspring about starting and continuing a pregnancy. I had also read some papers about the phenomenon of genomic imprinting (the phenomenon that the expression of alleles in an organism depends on the sex of the parent) and decided to discuss problems with Rolf Hoekstra, then a population geneticist at Wageningen University, because Hoekstra had studied conflicts between mitochondria and host cells and how they were resolved. He told me that the Australian biologist David Haig had already explored these problems. When I later read Haig's (now well-known) paper on 'Genetic conflicts in human pregnancy' (1993), doubts about the basic soundness of these ideas disappeared. About ten years later I visited Haig and discussed possible effects of intragenomic conflicts on brain and behavioural development. I am indebted to Haig for sharing his ideas on intragenomic conflicts. The discussions helped me to think more clearly about how imprinted genes affect the transfer of resources from parents to children.

Imprinted genes have key roles during prenatal development in resource transfer from mother to child. They are also expressed in the brains of children and then affect the postnatal transfer of resources from parent to offspring. They are also involved in the development of communicative behaviours, including linguistic behaviour. This latter observation raised the possibility that, if we will learn more about the effects of imprinted genes on brain development, then we will understand the social evolution of the human mind better. It requires no stretch of the imagination to understand why this possibility created excitement among investigators. There was another reason: studies by Keverne *et al.* (1996; see also Allen *et al.*, 1995) showed that paternally and maternally derived genes have different effects on the relative growth of brain structures. Paternally expressed genes contribute mainly to the development of the hypothalamus, while maternally expressed genes contribute to the development of the neocortex. Does this mean that genes that come from the father promote impulsive, instinctive behaviour, whereas genes that come from the mother promote more conscious, reasoned behaviour? Are these different effects also the reason why there are intrapersonal conflicts in the mind? Elucidating the effects of imprinted genes on brain development, many reasoned, will provide us for the first time with information on how and why genes affect – what Freud and James called – conflicts in the mind. For me there was another reason for excitement: I could use the alternative insights of Hacker and others to study the possible effects of imprinted genes on the development of the human mind. This possibility was the reason for a second wave of trips to Oxford, and I started to think about how we can integrate modern evolutionary theory and the philosophy of mind advocated by Hacker and others.

In this book I summarize the main results of my investigations into the social evolution of human nature. It describes how human nature evolved. It is up to readers to evaluate the way I have integrated modern evolutionary biology (i.e. inclusive fitness theory) and an alternative conception of human nature. I really enjoyed rethinking some problems in evolutionary biology and adjoining fields (medicine, psychology, anthropology) and coming up with new ways of studying them. It is easy to summarize the major shift advanced in this book: it is more interesting to study problems with the Aristotelian framework extended with Hamilton's theory than with the Cartesian framework extended with Hamilton's theory. The reasons why are explained in this book.

The thoughts in this book developed slowly. When I started the project of writing this book, I thought that it would take one or two years to finish it. It took me about five years, because solutions did not come to me as quickly as I hoped. But I enjoyed tackling the problems and it was worth the effort, for every step I took during this project was an improvement and helped me to see the social evolution of human nature better. I can only hope that readers will enjoy the picture sketched in this book and that it will help them to solve and resolve problems they encounter. I immediately add that this picture is based on the insights of others, especially on the ideas expressed by the persons mentioned in this preface. I want to thank them for sharing their ideas with me. I am especially indebted to Peter Hacker. I have benefited over a period of thirty years from his writings, comments and from our discussions.

A shortened version of Chapter 4 was published in *Biological Theory* (2010, 5: 357–365). Chapter 6 was published in *Studies in the History and Philosophy of Biology and the Biomedical Sciences* (2010, 41: 263–271) and a version of Chapter 5 was published in *Theory & Psychology* (2011, 21: 377–395).

CHAPTER I

The major evolutionary transitions and Homo loquens

1.1 Introduction

The biological world contains creatures such as fungi, plants, self-moving animals and self-conscious and culture-creating humans. These are composed of cells and cell organelles that were separate in the evolutionary past. Cell organelles were originally cells but later joined other cells (they became endosymbionts) and formed a higher-level creature: the symbiotic unicell (i.e. the eukaryotic cell). In this higher-level unit cell organelles retained their capacity to reproduce. Symbiotic unicells later joined other unicells and formed another higher-level unit: the multicellular organism. In this unit, the somatic cells abandoned their capacity to contribute genes to the next generation.

The benefit of cooperation is the reason why cells (and later multicellular organisms) joined forces (Bourke, 2011; Buss, 1987; Maynard Smith and Szathmáry, 1995; 1999). But there have been conflicts among the different cells as well. Conflict occurs when units are capable of affecting a common feature and when natural selection favours different effects of the units. For instance, the cells forming the symbiotic cell were unrelated and had therefore different interests, creating the potential for conflict. And in a large multicellular organism there were opportunities for cells to pursue their own ends, selfishly disregarding the common goals of the higher-level unit (e.g. a cancerous cell). Because selfish subunits undermined the stability of the new unit, the potential for conflict was reduced through conflict mediation. Conflict mediation led to subunits that no longer favoured opposing effects on a common feature, but a similar effect. The association between the subunits was therefore further strengthened. Hence cooperation and subsequent mediation of conflict contributed to an increase in life complexity: associations between lower-level units led to higher-level units, and reducing conflicts led to stronger associations (i.e. the integration and coordination of processes). As the result of reducing conflict the collectives were able to evolve further: they became obligate collectives with a high degree of interdependence of their parts. As obligate collectives, they became themselves candidates for participation, as subunits, in the next evolutionary transition (Bourke, 2011). The symbiotic unicell, multicellular organisms and later animal societies are therefore composed of layers upon layers of cooperation.

The evolution of higher-level units out of lower-level units is explicable in terms of the framework of inclusive fitness theory. It is no exaggeration to say that this insight is one of the most important theoretical discoveries of the previous decades. It was made possible by three developments. First, the discovery of a complex subcellular world. Molecular biologists disclosed a natural world below the level of the cell that did not rank second in richness to the world already known to naturalists. The discovery of cell organelles (e.g. mitochondria and chloroplasts), of genetic parasites such as transposons and retroviruses, of the processes of cell differentiation and apoptosis, of protein synthesis and RNA splicing and so on and so forth, revealed an enormous diversity of molecules and biological structures interacting in an orderly manner. Just as in the world studied by naturalists, some molecules and structures were related because of common descent whereas others were not. These observations raised the problem of how we can explain the orderly interactions between molecules and structures. Second, the development of a general theoretical model of social evolution. Darwin showed that natural selection is capable of explaining patterns observed by naturalists. Elaborations of Hamilton's theory revealed that inclusive fitness theory has similar explanatory powers and is capable of explaining observable patterns in both the subcellular and supracellular world. Hence the same theoretical principles that explained the behaviour of organisms and the social life of animal societies, could also explain the behaviour of genes, cell organelles and cells. Third, the development of a coherent conceptual framework for understanding social evolution. Scientists and philosophers have elaborated the rules for the use of both technical and ordinary concepts that we use when we investigate the behaviour of humans and other organisms. These investigations revealed that mastery of a language is the mark of the human, rational mind. This conceptual insight raises the question of how linguistic behaviour evolved out of animal behaviour. Hence the challenge is to explain how *Homo loquens* evolved in terms of Hamilton's theory. The principles and concepts of a framework capable of explaining the social evolution of human nature are discussed and elaborated in this book.

Table 1.1: Two kinds of major evolutionary transitions, adapted from Queller (1997, 2000).

	Egalitarian	Fraternal
Examples of cooperative alliances forged	Different molecules in compartments; genes in chromosomes; nucleus and organelles in cells; individuals in sexual unions	Same molecules in compartments; same organelles in cells; cells in individuals; individuals in colonies
Units	Unlike, non-fungible	Like, fungible
Reproductive division of labour	No	Yes
Control of conflicts	Fairness in reproduction; mutual dependence	Kinship
Initial advantage	Division of labour; combination of functions	Economies of scale; later division of labour
Means of increase in complexity	Symbiosis	Epigenesis
Greatest hurdle	Control of conflicts	Initial advantage

1.2 The major evolutionary transitions

The evolution of a higher-level unit out of lower-level units is called an evolutionary transition. The lower-level units constituting the new unit lack a high degree of conflict within the new unit and interact in a coordinated manner to achieve common goals. I shall later elaborate this general definition of an evolutionary transition, but first discuss two types. Queller (1997, 2000) subdivided the major transitions into *egalitarian* and *fraternal* transitions (see Table 1.1). He took these terms from the motto of the French revolution: 'Liberté, égalité, fraternité'. The essential difference is that egalitarian transitions involve a union of unrelated units, whereas fraternal transitions involve related units. Units involved in an egalitarian transition do not sacrifice their reproductive capacities when the units cooperate (hence their egalitarianism). The evolution of the symbiotic cell is the paradigmatic example. Other examples are the grouping of unrelated genes in chromosomes and the union of the female and male halves in a sexual organism. The benefit of cooperation is that two different functions are combined. The fraternal transition, by contrast, consists of a union of related units (hence their fraternity). The evolution of a multicellular organism is here the paradigmatic example. Because the cells are related (identical if the organism is derived from a single cell), there are no benefits related to combining different functions. The initial advantage was probably related to an increase in size. When multicellular organisms started to compete with other organisms, selection favoured a division of labour within the organism, resulting in cells and organs performing different functions (as the result of epigenesis, i.e. through different gene expression different types of cells, tissues and organs evolved). An essential division of labour was between germ cells and somatic cells. Somatic cells are characterized by sacrificing their reproductive potential. Other examples of fraternal transitions are the cooperation of individuals in a colony and the grouping of the same molecules in a cellular compartment.

This distinction between fraternal and egalitarian transitions immediately explains why the transition from unicellular to multicellular organisms and from cells (prokaryotes) to symbiotic unicells (eukaryotes), faced different problems. In the transition to multicellular organisms *kinship* explains why lower-level units cooperated and why in small multicellular organisms conflict was absent, whereas during the transition from unicells to symbiotic unicells conflict between unrelated cells was the greatest hurdle. But since endosymbiosis combined unrelated entities capable of performing different functions, *mutual benefit* explains why the new entity had adaptive advantages. By contrast, the lack of a combination of different functions was the greatest hurdle for a multicellular organism.

The distinction between egalitarian and fraternal transition also clarifies why certain phenomena occur in only one transition. For instance, an egalitarian transition, in contrast to fraternal transitions, does not involve *altruism*, for altruism, according to the principles of social evolution, can only occur if the interacting units are related. This amounts to the earlier observation that sacrificing the ability to reproduce (e.g. the non-reproductive, somatic cells of a multicellular organism and the sterile workers of social insect societies) is absent in egalitarian transitions. The cooperation between endosymbionts and host cells is therefore only explicable in terms of mutual benefit (also called 'narrow-sense cooperation'): both partners benefit because of the interactions and there is no indirect benefit (because of their relatedness) involved. Note that the framework of inclusive fitness theory is not restricted to explaining interactions between related individuals. It is also capable of explaining patterns of interactions between unrelated subunits, but uses then different principles. Hence inclusive fitness theory encompasses both kin selection theory (explaining cooperation among related individuals) and theories that explain cooperation between non-relatives.

Table 1.2: Six major evolutionary transitions, adapted and modified from Bourke (2011).

	Social group formation	Social group maintenance	Social group transformation
1. Separate replicators (genes)→ cell enclosing genome	Origin of compartmentalized genomes	Control of selfish DNA	Evolution of large, complex genomes
2. Separate unicells→ symbiotic unicell	Origin of eukaryotic cells	Control of organellar reproduction	Evolution of hybrid genomes through transfer of genes from organelles to nucleus
3. Asexual unicells→ sexual unicells	Origin of zygotes	Control of meiotic drive	Evolution of obligate sexual reproduction
4. Unicells→ multicellular organism	Origin of multicellular organisms	Control of selfish cell lineages (cancers)	Evolution of a segregated, early-diverging germ line
5. Multicellular organisms→ eusocial society	Origin of societies	Control of conflict with dominance, punishment or policing	Evolution of dimorphic reproductive and non-reproductive castes
6. Primate societies→ human societies	Origin of language	Control of cheating	Reciprocating, evolution of social norms

I distinguish here six major evolutionary transitions (see Table 1.2): (1) the evolution of genomes, (2) the evolution of eukaryotic cells, (3) the evolution of sexual reproduction, (4) the evolution of multicellular organisms, (5) the evolution of societies, and (6) the evolution of human societies (language). Maynard Smith and Szathmáry (1995; 1999) defined eight transitions. They subdivided the first transition in Table 1.2 into three separate transitions. Their table has been criticized and modified by Bourke (2011) whom I am following here. But whereas Bourke defines the evolution of the origin of interspecific mutualism (cooperation between separate species) as the sixth transition, I return to the evolution of language defined by Maynard Smith and Szathmáry as the, for the time being, final transition. There is only one reason: this book discusses the problem of how we can understand the transition from primate to human societies. Bourke focuses on another no less interesting problem, but this problem is not under discussion here. However, he has an important argument for excluding language evolution: it does involve the evolution of – what he calls – individuality, i.e. a new biological entity. I include language evolution here because it enables humans to create new bonds (reciprocity, social norms) resulting in new collectives consisting of individuals interacting in a coordinated manner to achieve common goals. One can argue that language evolution did not result in a transition but in a transformation of primate societies. This would meet the objection that the resulting collectives are not strict biological entities, but has the disadvantage that language evolution and its consequences are not comparable to the many transformations in the animate world (e.g. the metamorphosis of a caterpillar into a butterfly). Language evolution created new (types of) bonds between organisms and led to cultural evolution: symbolic information was transmitted from one individual to the other. The problem discussed in this book is of how we can expand the framework of inclusive fitness theory so that it is capable of explaining the sixth transition.

However, it is important to note that speaking about language evolution as an example of an evolutionary transition does not mean that human societies are, or are predicted to become, biological entities (as Bourke correctly noticed). This has the important consequence that human societies are not and will not evolve into *obligate* collectives. In obligate social groups the lower-level entities can only replicate as part of the group. For example, in the case of multicellular organisms, it means that cells in obligate multicellular organisms can only replicate as part of an organism, whereas in *facultative* multicellular organisms (e.g., in social amoeba, see section 1.3) they have the capacity to replicate independent of the higher-level entity. There is evidence that obligate social groups evolve when $r=1$, for there are no examples known of obligate social groups when $r<1$ (Fisher, Cornwallis and West, 2013), i.e. when social groups are formed out of an aggregation of cells that are not always identical. Interestingly, there is a similar story to tell in the case of insect societies: obligate insect societies (with sterile females) only evolve when there is monogamy (Boomsma, 2009), leading to a potential worker being equally related to her own offspring and to the offspring of her mother ($r=\frac{1}{2}$ in both cases; the relation to the offspring of her mother is the average of three-quarters (to the daughter of the mother) and one-quarter (to the son of the mother)). Any small efficiency benefit for rearing siblings (offspring of the mother) over their own offspring will then favour eusociality (i.e. a division of labour resulting in an obligate insect society). Hence it appears that a bottleneck is essential for the formation of obligate collectives: they evolve when a multicellular organism is derived from a single cell, and when an insect society is started by a single queen fertilized by a single male.

Yet although current human societies are not obligate societies, social groups evolved. One can argue that the stability of human societies requires a social-cultural explanation, but that does not exclude inclusive fitness theoretical explanations. I shall argue in this book that inclusive fitness theory applies here because the use of a language evolved as an extension of non-verbal, communicative behaviour. But I shall also argue that to use cultural and historical explanations becomes more and more important when humans are able to use a complex language.

The six major evolutionary transitions took time (see Bourke, 2011). It is estimated that the first cell arose around 3,500 million years ago (mya). Many events must have occurred between the origin of self-replicating molecules and the origin of the first cell. For example, there has been a switch from RNA to DNA as the primary replicator, and the genetic code and protein synthesis evolved during this period. The first symbiotic unicell arose 2,000 mya. The eukaryotic cell was 1,000 times larger by volume than its prokaryotic predecessors. It acquired not only cell organelles (mitochondria, in the later animals, and chloroplasts, in the later plants) as the result of the symbiotic fusion of two cells, but also possesses several structures which are absent in prokaryotic cells, such as a nucleus and internal cytoskeleton. It is unclear when the transition from asexual unicells to the sexual unicell occurred, but it can be placed between the origin of eukaryotes (2,000 mya) and the origin of multicellular organisms (1,200 mya), because it has not preceded the origin of eukaryotes. Prokaryotes (e.g. bacteria) transfer genetic material to one another, but sexual reproduction in eukaryotes involves the formation of haploid gametes from a diploid cell and the fusion of the gametes to form a zygote. Multicellular organisms arose 1,200 mya and there is evidence that multicellularity evolved at least sixteen times independently in eukaryotes. Simple multicellular eukaryotes are slime moulds (social amoeba); complex multicellular eukaryotes are animals and plants. The first eusocial societies (a social system with non-reproductive workers) consisting of multicellular organisms arose 150 mya; the first steps resulting in language evolution were taken about 2 million to 3 million years ago.

Bourke (2011) subdivided the stages of evolutionary transitions into three principle stages (see Table 1.2). First, social group *formation*: i.e. the initial formation of a higher-level unit as the result of the spread of genes for social behaviour through the population. Kinship and mutual benefit explain why units join forces. Second, social group *maintenance*:

i.e. the stable persistence of higher-level units once they have originated (e.g. through mediation of potential conflicts between subunits). Because the stability of the new-formed unit depends on the coincidence of fitness interests, fraternal unions are maintained through self-limitation, whereas egalitarian unions are maintained through shared reproductive fate. A well-known example of the latter principle is the uniparental inheritance of mitochondria: they are transmitted through the female gamete reducing the potential of conflict. For if mitochondria are transmitted only through the female germ line, a mitochondrial variation occurring in the male will not be passed on to the next generation, and the success of a mitochondrial variation arising in the female depends on the success of the female (see further in Chapter 3). Third, social group *transformation*: i.e. the process that transforms the facultative higher-level unit into an obligate one. For example, in a simple, facultative multicellular organism, somatic cells are totipotent (in insect societies workers have reproductive potential), whereas in a complex, obligate multicellular organism all somatic cells lost totipotency and the organism displays a segregate, early-diverging germ line (in insect societies workers have low reproductive potential and the society displays high queen–worker dimorphism).

Units coordinate their actions in order to achieve common goals. Inclusive fitness theory explains these goals in evolutionary terms, i.e. goals are defined in terms of reproductive fitness. I shall later explain why another interpretation of 'common goals' is essential for understanding the transition to human societies, but first discuss the evolutionary view. The advantage of the evolutionary definition is that it enables us to see the common principles of social evolution. For example, compare a multicellular organism consisting of interacting cells with an insect society consisting of interacting organisms. In both cases kin selection explains why new levels of social organisation evolved consisting of lower-level units (cells or organisms) coordinating their activities or actions so that fitness is maximized. There is, of course, an important difference: in the case of multicellular organisms the cells are identical ($r=1$ if there are no mutations) if they are derived from a single cell; in the case of an insect society the relatedness between sisters is three-quarters if the society is founded by a single queen fertilized by a single male. Because males develop out of unfertilized and females out of fertilized egg cells, the sisters share the common genome of their father but only one of the genomes of their mother; hence the relatedness is $(1+\frac{1}{2})/2=\frac{3}{4}$. Yet kin selection explains why cells coordinate their actions and why females are the cooperative part of the insect society, and why somatic cells and workers sacrificed their reproductive potential. Kin selection theory also predicts that both transitions are contingent upon suppression of conflict between lower-level units. One can predict that conflict was more abundant during the transition to insect societies because of the lower value of r (see Ratnieks, Wenseleers and Foster, 2006).

Hence in both cases kin selection explains the evolution of a new entity consisting of lower-level entities interacting to maximize reproductive fitness, either as parts physically joined to one another (a multicellular organism), or as parts that remain and tend to remain in close proximity (an insect society). Note that only humans can create bonds between individuals even if they are not living in close proximity to each other. Note also that there is no separation between altruistic and reproductive functions in human societies, and that language evolution enabled humans to form intentions and, hence, to choose between selfish and altruistic behavioural options. It is, therefore, a challenge to extend the conceptual framework of inclusive fitness theory so that this extension (1) highlights the common principles of social evolution and (2) is capable of accommodating the transition made possible by language evolution.

Fraternal and egalitarian unions are two extremes and it is possible to characterize the major transitions, as described in Table 1.2, as fraternal, egalitarian, or both. For example, the evolution of workers in social insects is a characteristic of a fraternal union, while the union of the female and male halves in the genome of sexual organisms is an example of an egalitarian union. The evolution of genomes probably involved both egalitarian and fraternal elements, because it included identical genes and unrelated ones.

1.3 Cooperation and conflict

Inclusive fitness theory solved the problem of altruism through invoking indirect fitness effects. For example, a cell of the immune system displays 'suicidal altruism' (apoptosis) when the cell is infected by a virus. It exhibits this behaviour because it promotes the survival and reproduction of its relatives in the whole organism. For the effect of apoptosis is not only that the cell dies, but also that the virus present in the cell is degraded by enzymes. Hence apoptosis is part of a defence mechanism: it reduces the survival and replication of the virus and enhances therefore the survival and replication of the genes (of the cell committing suicide) because copies of those genes are also present in the germ line of the organism. Suicidal altruism is therefore explicable in terms of inclusive fitness theory because of its indirect fitness effects.

Inclusive fitness theory teaches us that the ultimate goal of genes is to maximize their transmission to future generations. This can be calculated and modelled, for this definition of the (ultimate) goal can be described as the effect of genes on the (expected) offspring numbers. Evolutionary biologists are not concerned with the goals or intentions of an organism when they study goals: they only investigate fitness effects. This clarifies why they also attribute goals to cells and talk about the 'suicidal altruistic behaviour' of the cytotoxic T-cells of the immune system (see further in Chapter 3). Note also that when they talk about 'selfish genes', the resulting behaviour displayed by cells or organisms may be either altruistic or selfish. I shall later elaborate this view, but first discuss three well-known examples illustrating how inclusive fitness theory explains cooperation and conflict at the subcellular level (see Burt and Trivers, 2006; Keller, 1999; Stearns and Hoekstra, 2005, chapter 9).

1.3.1   *Transmission of bacterial plasmids*

Many bacteria carry circular DNA molecules called plasmids. These have genes important for their own propagation, but they also contain genes which are beneficial for the bacterial cell, like those that code for antibiotic resistance. Plasmids are transferred vertically, i.e. at bacterial cell division they are transmitted to the daughter cells. If this was the only possibility, then their long-term fate would be coupled to that of the bacterial host: the more successful the host, the more successful plasmids are. Inclusive fitness theory explains then why the possible actions of plasmid genes coincide with those of the host. However, horizontal transmission occurs too. The genes of many plasmids are able to induce their host to conjugate with another, uninfected cell. During conjugation the recipient acquires a plasmid while the donor retains a copy. The consequence of horizontal transmission is that the long-term fate of plasmids is no longer coupled to the success of the host cell. Models show that, if the horizontal transfer occurs frequently enough, plasmids can invade and establish themselves despite a negative effect on the bacterial host. Hence horizontal transmission creates the potential for conflict between the host and plasmids. If plasmids have negative effects, natural selection at the level of the host will favour the loss of plasmids, since plasmid-free bacteria do not bear the costs. Natural selection also predicts then the evolution of counteractions of plasmids.

Plasmids have evolved a mechanism preventing them from being removed from the bacterial cell. They make the host addicted to their presence by producing both a toxin and its antidote. The gene producing the toxin and the gene producing the antidote belong to a team (however, the antidote is less stable than the toxin). If the plasmid is present, then there is no problem, because this guarantees that both toxin and antidote are present. However, when the plasmid is not transmitted to the daughter cell, the cell dies because the antidote has disappeared while the toxin is still present. Note that the addiction mechanism is only selected for in the case of horizontal transmission.

1.3.2 Slime moulds and cooperation

An example illustrating cooperation and conflict in a simple multicellular unit is described in the social amoeba *Dictyostelium discoideum*. This creature is a solitary cell for most of its existence. The cell moves through the soil with pseudopods, engulfs bacteria for food and periodically divides by mitosis. However as the result of starvation the transition to multi-cellularity is triggered. If there are a sufficient number of amoebas in the area, they aggregate and form a multicellular slug that crawls along towards light and heat (and away from ammonia). These cells form a fruiting body. In this body some cells form a stalk lifting other cells above the substrate. These other cells form spores awaiting dispersal to a better environment. The point to notice is that the multicellular stage has the potential for cooperation but also for conflict. For spore cells contribute to the next generation, whereas cells that contribute to the stalk do not. Consequently cells that produce spore cells but not stalk cells have an advantage. These are called cheaters. One way to avoid the risk of cheaters is kinship: if the cells that produce the stalk and those that form the spore cells (of the fruiting body) are related, then their interests are the same. This is often the case since the multicellular organism is formed from cells in a local population. Hence limited dispersal explains cooperation and why cheaters will be held in check because of the costs imposed by selfishness (self-limitation in the fraternal group). However, it is also possible that the multicellular organism is composed of unrelated cells enhancing the potential for conflict. There is evidence that a genetic recognition mech-anism evolved to reduce the risk of cheaters in this case (Queller *et al.*, 2003): only cells with a certain adhesion gene (coding for a product needed for cells to adhere to each other) form a fruiting body, excluding cells which are recognized as having a different adhesion gene.

1.3.3   X and Y chromosomes

Conflicts can also occur if genes, or teams of genes, affecting a trait follow different transmission rules. Examples are paternally and maternally inherited genes, since the probability that they are present in offspring may be different, and genes on the sex chromosomes, which follow different transmission rules compared to genes on the autosomal chromosomes. I discuss here the sex chromosomes.

Suppose that there is a gene on the Y chromosome that has the capacity to produce a substance that somehow kills germ cells carrying an X chromosome (see Hamilton, 1967; Ridley and Grafen, 1981). If this gene is present in a male, he will produce only sperm cells with a Y chromosome. It is easy to see that such a gene will be favoured if the reduction of the number of eggs sired by the male is no more than 49%. For in that case more than 51% of the fertilized eggs have the Y chromosome, while in the absence of this Killer gene on the Y chromosome only 50% of fertilized egg cells would have the Y chromosome. This explains why a Killer gene occurs. But such Killer genes are the exception rather than the rule, and to see why, we have to discuss the effects of the gene on genes on the autosomal chromosomes. The Killer gene is not only a disadvantage for the genes on the X chromosomes, but also for the genes of the autosomal chromosomes, for in the model discussed above, their contribution to the next generation is reduced by up to 49%. Inclusive fitness theory expects the evolution of modifier genes that suppress the action of the Killer gene. And since there are more genes on the autosomal than on the sex chromosomes, autosomal genes have more power in the 'parliament of genes' (Leigh, 1977).

Inclusive fitness theory clarifies why the genes located on autosomal chromosomes cooperate. Assume that the opportunities for competition or cheating are limited. Then genes that are part of a group can increase their own success by increasing the success of the whole genome, i.e. their group (Frank, 2003). They receive then their fair share of this success. Consequently, any mechanism that aligns reproductive interests or represses competition within groups will then select for higher levels of cooperation. Fair meiosis may be an example of such a mechanism and appears to have been favoured because it aligns the reproductive interests of genes in a genome. For under the rules of fair Mendelian transmission, every gene in the genome has an equal probability of being passed on to the individual's offspring, so it is in the interests of all genes to maximize the reproductive success of the individual.

Haig and Grafen (1991; see also Ridley, 2000) have argued that, in general, the evolution of segregation distorters (i.e. genes that distort the fair distribution of genes among the gametes) was further reduced by genetic recombination; for recombination is a general mechanism for reducing the advantages of segregation distorters such as the Killer gene. In order to see why, we have to think about the target of the Killer gene. Until now I have said that Killer is capable of killing cells with the X chromosome. But how does Killer 'know' the difference between cells with the X and Y chromosome? Besides a killing-mechanism, there has to be a recognition mechanism such that Killer only kills cells with an X chromosome. Thus Killer must act together with a Recognition gene that is capable of recognizing a Target (a certain sequence of nucleotides) of the X-bearing germ cells, with the effect that Killer does not attack Non-target Y-bearing germ cells.

Recombination is a mechanism that has the capacity to decouple the association between the Killer gene and the Recognition gene (and Target and Non-target). The consequence of recombination is that the Killer is invited to play Russian roulette: there is a possibility that, after recombination, Killer attacks a Non-target and, hence, kills itself. Recombination is for this reason thought to be a general mechanism preventing the evolution of distorters that disturb the spread of genes according to the principles of the fair laws of Mendel.

1.4   The evolutionary gene: what is it?

These examples illustrate why evolutionary theorists talk about the social behaviour of genes. Yet what are genes and how do they affect processes at different levels of organization? First, they affect processes because genes are *expressed*. Genes consist of a particular sequence of nucleotides (a DNA sequence) which can be transcribed into mRNA. Messenger RNA can have a function in a cell as a regulator of processes, and can be translated into a protein (which may function as an enzyme, hormone, etc.). Second, genes affect processes because they are *replicated*: as the result of DNA replication, copies of the gene are produced. Replication can be within the somatic cells of an organism (as part of cell divisions resulting in, for example, growth of an organism), and within the germ line of an organism (as part of the reproduction of the organism). Although gene expression and gene replication are two different processes, molecular genetics has taught us that the same information (the genetic code) is involved in both processes.

Genes have social effects. Genes of plasmids cause their host cell to conjugate with another cell; genes in the female bees of royal court cause them to care for the queen, reducing the chance that she will die of infectious diseases; and genes in mammals cause self-grooming and allo-grooming reducing the spread of ectoparasites in the group. Inclusive fitness theory explains why genes have social effects. Dawkins (1976; 1982) famously made a distinction here between the gene and its vehicle. In his view genes are always 'selfish', but the vehicles of genes can be selfish or altruistic. The reason is that cooperation at, for example, the organismal level has its basis in the 'selfishness' at the level of the genes. For example, when an organism displays altruistic behaviour, then Dawkins argues that it does so because a gene for altruism is levering its way into the next generation through promoting aid to relatives that bear the same gene. I shall elaborate the gene's eye view of evolution through discussing the distinction between the gene and its vehicles.

In order to explain how genes affect processes in their vehicles, it is useful to distinguish the informational gene and the material gene. When DNA replicates, the gene consisting of a particular DNA sequence is replicated too. The original gene is then, as a result of the replication, replaced by two genes. The more often DNA replicates, the larger the number of genes. But as long as there are no mutations, the sequence remains the same, no matter how many times DNA replicates. Thus it makes sense to differentiate the *informational* gene (the particular DNA sequence that remains the same) and the *material* genes (the copies of the gene present in, for example, a cell or organism). The distinction between the 'selfish' gene and its vehicles corresponds to the distinction between the informational and material gene. But there is more to say, for this distinction raises the question of how the informational gene affects processes (see Haig, 1997).

Consider a new informational gene that evolved as the result of a mutation. If it is a successful gene, then it must be able to invade and maintain itself as a rare variant, for there is, at first, only one copy. If the gene acts in isolation, then the material copies of the informational gene must promote their transmission in the cell, organism or society in which it is expressed, in order that the informational genes survives. The effects of the material gene are then crucial for the success of the informational gene. Now suppose that the gene is present in an organism with a germ-soma division, then the effects of the material gene in the somatic cells do not contribute directly to the transmission of the informational gene to the next generation. But since there are copies of the informational genes present in the germ cells, the effects of material genes may contribute indirectly to its transmission to the next generation.

In this scenario, the success of a new informational gene is relative to an already present gene in the gene pool. For the new informational gene evolved as a mutation of an existing gene. In other words, the success of a new gene depends on its capacity to replace the already existing variant. If we conceive of genes in this way, then they can be investigated with the aid of models from game theory. Just as game theorists study behavioural strategies and ask whether a new, mutant behavioural strategy is able to invade a population of individuals displaying an alternative strategy, we can ask whether a new gene is able to invade a gene pool in which an alternative gene is present. For if we compare different genes (for example, at one locus), there may be differences in their strategic effects, for example, in the number of material genes that they produce or in the number and effects of the transcripts they produce. The usefulness of game theoretical models for studying genes depends on the choice of the set of alternative strategies chosen (the patterns of gene expression), and on the possibility to test models with the help of data obtained in empirical studies.

In order to make this abstract explanation more concrete, I discuss two extreme examples: the effects of genes located on a plasmid, and the effects of genes located on a chromosome of a human. Suppose that we want to use game theory for understanding why a gene located on the circular genome of a plasmid causes its host to conjugate with another cell (and, hence, creates the possibility for horizontal transmission). Then we can ask whether this gene can invade a gene pool in which there is another gene only contributing to the vertical transmission of plasmids. These investigations tell us that such a gene is able to invade the gene pool. If the gene causing horizontal transmission has negative side effects on the host, then we can calculate what the gain in fitness has to be, compared to the loss of fitness as the result of the side effects. Suppose that we want an evolutionary answer to the question why in human populations some males display violent, impulsive behaviour while others exercise self-control (or why one individual is capable of displaying both strategies in different circumstances). Game theory enables us to develop a first answer. Assume that there is a population in which all young males execute self-control so that conflicts are always solved in a peaceful manner (this is a hypothetical situation). What circumstances favour the invasion of the aggressive variant that can be established in this population? Perhaps a large gap between rich and poor is an example of such a circumstance because the new variant has initially a higher reproductive success than the existing variant. Calculations may show that in the end a stable equilibrium will evolve consisting of a certain ratio of the self-control and impulsive strategies. Note that the hypothetical differences in the investigations of the game theorists are related to alternative worlds which do not necessarily need to exist: it may be an imaginary world, a past world that no longer exists, or a future world.

Informational genes have social effects through material genes, cell organelles, cells and so forth, and these are modelled with game theory. In order to investigate these effects, Haig has argued that we should describe the effects of the informational gene through the material genes and their transcripts (and the effects at different levels of organization) by the term *strategic gene* (Haig, 1997; 2012). Genes are called 'strategists' because their effects can be studied with game theory (and other models developed in the field of evolutionary genetics).

The gene's-eye view studies strategic genes at the level of populations. Because of their effects on fitness certain alleles are selected. Evolutionary geneticists differentiate here the level of selection and the level of adaptation. Genes are expressed at different levels of organization: they are the genes of cells, organisms and societies. Yet genes are special because they are transmitted to the next generation (not the different vehicles); they rise and fall in frequency; and they have primacy over the vehicles in life cycles because genes 'construct' vehicles and 'cause' them to act. This matters because adaptations serve genes, not the different vehicles. Vehicles can act selfishly or altruistically, yet both acts are at the level of the gene the effect of a strategic gene. Thus, there are different levels of selection, but the gene-view is a way of highlighting that genes are exceptional since they are the *units of adaptation*. Selection is at the level of the phenotype (the different vehicles and their actions), but what is passed on to the next generation are not phenotypes but the corresponding genotypes.

Calling genes strategic is a way to highlight the social effects of genes. Molecular biology revealed that genes affect processes in cells and organisms in many different ways. Genes are expressed in some tissues and organs, but not in others; they have effects during specific developmental stages; some genes operate in teams achieving a common goal, whereas others act on their own; they oppose the effects of other teams or cooperate with them; they respond to certain signals from the environment and from other genes; they have as it were a memory, because they are expressed differently when maternally or paternally derived; they have different transcripts (due to splicing) and use these in different contexts; and so on and so forth (Haig, 1997, p. 285). The discovery of the subcellular world is a major reason why investigators started to talk about the social actions of genes. For their 'social effects' were not captured by the traditional vocabulary of Mendelian genetics: genes are either dominant or recessive. Assigning agency to genes is a way to reveal all the possible expression patterns that genes use to maximize their transmission to the next generation.

1.5   The causation of phenomena

The gene's-eye view has raised many questions. It is not my intention to solve or resolve them once and for all. I only discuss here three important reasons why evolutionary theorists adhere to the gene's view of evolution.

The *first* reason why evolutionary theorists endorse the gene's-eye view of evolution is that it is backed up by Fisher's theorem of natural selection (see Box 1.1). In the modern version, first elaborated by Price (1972; see also Okasha, 2008), this theorem states that the increase in average fitness of the population ascribable to a change in gene frequency is equal to the additive genetic variance in fitness. The additive genetic variance is the fitness variation in the population that is due to additive (or independent) action of the genes. The theorem does not deny but excludes non-additive genetic variance, i.e. variance which is brought about by interactions between genes (also called the epistatic component of variance). The reason why Fisher excluded non-additive variance is that a gene's effect on fitness depends in the case of non-additive variance on other genes: it is dependent on genetic background. And genetic background changes every generation because genes are reshuffled during sexual reproduction generating novel combinations. Consequently, the fitness effect of a gene may change too. For example, it may result in higher or lower variance in fitness because of the change in genetic background. Hence in contrast to additive genetic differences that are heritable, non-additive genetic differences are not heritable, and this is the reason why Fisher excluded non-additive variance. The non-additive variance belongs in Fisher's model to the environment. The environment is in Fisher's model a broad category including climate and interactions with other species, but also non-additive causes of changes of gene frequency such as epistasis.

Fisher's model shows that natural selection leads to an increase in average fitness if selection operates on the additive genetic variance in a population. Hence it appears that natural selection is directional. Fisher suggested that his theorem is therefore comparable to the second law

Box 1.1: The Price equation and Fisher's theorem

Consider a population containing n entities indexed by i. Assume that these entities are individuals, but realize that they can also be genes, cells or groups. Let w^i be the absolute fitness of the ith entity, i.e. the number of offspring it contributes to the next generation. The relative fitness is then $\frac{w_i}{\overline{w}}$ (overbars denote population average). Since we are interested in the effects of selection on a character, this character is denoted as z. The value of the character is z_i for the ith parent and z'_i for the offspring (the prime is added to z here to denote the value of z in the next generation). The change in character occurring through transmission from parent to offspring is Δz_i, whereby $\Delta z_i = z'_i - z_i$. Price demonstrated that the average change in the value of the character $z(\Delta\overline{z})$ from one generation to the next is given by:

$$\Delta\overline{z} = \text{cov}\left(\frac{w_i}{\overline{w}}, z_i\right) + E\left(\frac{w_i}{\overline{w}}\Delta z_i\right) \qquad (1.1.1)$$

The first term on the right-hand side of the Price equation is the statistical covariance between the value of the character z_i and its relative fitness $\frac{w_i}{\overline{w}}$. It represents selection acting on the character from the first generation to the next. If the covariance is positive, then individuals with larger values of the character tend to have more offspring. For example, if z were height, then the average height will increase in the population on condition that individuals with a larger value of height have higher fitness. The second term represents the transmission of the character: it describes whether offspring differ from their parents. It is denoted by E, or expectation, taken across the population (arithmetic average). Expectation refers here to the expected value that would have resulted if the individual left offspring and describes the change due to transmission. Hence if it is zero (e.g. if in diploid organisms meiosis is fair), then there is no transmission effect. Transmission effects occur as the result of, for example, mutation, segregation distorters, recombination, or a change in the environment.

The Price equation is special for at least three reasons. First, because of its generality: it applies to any group of entities undergoing change and these entities do not necessarily need to be biological entities. Second, because it separates change due to selection and change due to transmission (the two components of the equation). Third, because the equation can be used as a meta-theory, for we can derive several relevant equations in the field of evolutionary genetics (for example, Hamilton's rule) from this equation. Therefore the Price equation can be used as a mathematical foundation for social evolution, multilevel evolution and cultural evolution, and helps mathematical orientated biologists to compare and contrast different models (see also Box 3.1 and 3.2). I briefly discuss here how we can use the Price equation in the context of population genetics.

In order to understand how we can use the Price equation for understanding changes in allele or gene frequencies, it is useful to recall that a covariance is equal to a variance times the regression coefficient:

$$\text{cov}(x, y) = \text{var}(x) \frac{\text{cov}(x, y)}{\text{var}(x)} = \text{var}(x)\beta(x, y) \qquad (1.1.2)$$

In population genetic models changes in gene frequencies are described, so we have to apply the equation to these changes. Price reformulated the equation by limiting the change in gene frequency to the additive genetic component (g) of a character. Hence he ignored the genetic change due to transmission (the second right-hand term in the original equation) and applied the equation to the change in average value of the character Δz. The equation results then in the following equation describing natural selection:

$$\Delta \overline{g} = \text{cov}\left(\frac{w}{\overline{w}}, g\right) = \beta\left(\frac{w}{\overline{w}}, g\right) \text{var}(g) \qquad (1.1.3)$$

This equation states that natural selection operates when there are heritable differences between individuals with respect to a certain character if this character correlates with reproductive success. Natural selection is here expressed as the product of two components: the regression slope of the relative reproductive fitness against the genetic value of the individual $\left(\beta\left(\frac{w}{\overline{w}}, g\right)\right)$, and the genetic variation in the population (var(g)). This summarizes the insight that natural selection operates whenever there are heritable differences between individuals and if the character correlates with reproductive success. Because variances are never negative, the equation states that the effect of selection must be an increase in reproductive success.

We can use this equation to derive Fisher's fundamental theorem (see Chapter 1, section 1.5). This theorem states that the change in the mean fitness of the population under the action of natural selection is proportional to the variance in fitness. Price explained that Fisher's theorem was a partial result, i.e. a description of the action of the natural selection effect when other effects are excluded. If we take fitness as the character ($z \equiv w$) and notice that the covariance of a variable with itself is its variance, then Fisher's theorem is described by:

$$\Delta \overline{w} = \text{cov}\left(\frac{w}{\overline{w}}, w\right) = \frac{\text{var}(w)}{\overline{w}} \qquad (1.1.4)$$

FURTHER READING

Gardner, A. (2008) The Price equation. *Current Biology* 18: R198–R202.
McElreath, R. and Boyd, R. (2007) *Mathematical models of social evolution: a guide for the perplexed.* University of Chicago Press.
Price, G. R. (1970) Selection and covariance. *Nature* 227: 520–521.
   (1972) Extension of covariance selection mathematics. *Annals of Human Genetics* 35: 485–490.

of thermodynamics. This law states that a closed system will always go from low entropy states to high entropy states. Yet commentators have noted that there is an important difference between Fisher's theorem and the second law of thermodynamics. Whereas the second law always applies, Fisher's theorem requires a qualification: the theorem only holds in a constant environment. The theorem is false when the environmental factors are not held constant. For example, the population may go extinct as the result of a drastic change in environmental conditions, and this is not explicable as an increase in average fitness. But if the environment is held constant, then the average effect of the alleles in the population is constant too. One can demonstrate that Fisher's theorem is then correct.

Why is Fisher's theorem an argument in favour of the gene's-eye view? In Fisher's model, the consequences of natural selection on the *genetic composition* in a population are studied. Because Fisher's theorem does not depend on the interactions between alleles (in a genotype), it describes the changes in allele frequency. Thus we can treat the population as a population of genes without paying attention to organisms. And that is the core of the gene's-eye viewpoint: we can understand natural selection by studying why effects of genes are selected.

The *second* reason is derived from molecular genetics: genes are essential in the evolutionary process because they are carriers of information. Molecular genetics has taught us that we can describe the sequence of the nucleotides in terms of the four bases characterizing the four building blocks of DNA: A, T, C and G. The point to notice is that four different nucleotides can generate an enormous amount of genetic variation. Consider a sequence consisting of one hundred nucleotides: there are then 4^{100} variations possible. In theory, each variation can result in an adaptation. These adaptations enhance fitness, i.e. the effect of a particular variant enables the carrier of the variant to survive in a specific environment. And because the genetic variation (a particular sequence of nucleotides) increases fitness in a specific environment through its effect, and since the effect has the result that the organism is adapted to the environment, one can argue that it has 'phylogenetically acquired information'. Because a particular sequence of nucleotides (which we can describe in terms of the four letters) has proven to be successful, there is as it were information 'stored' in the DNA as the result of natural selection because the specific DNA sequence is selected. Yet it is clear why this picture of storing information in the genome raises conceptual problems, for what is meant by information here? Some of these problems are discussed in Chapter 6.

The *third* reason is the distinction between *proximate* and *ultimate* explanations. This distinction concerns the causation of phenomena (Mayr, 1961; Tinbergen, 1963). It can be explained as the difference between answers to the *how* and *why* questions. Proximate explanations answer the question how behaviour develops (ontogenesis) and how it is caused by internal and external factors (the mechanisms of behaviour). It gives answers to questions such as: how does behaviour develop during the lifetime of an individual? Ultimate explanations answer the question why certain behaviour exists and what its contribution is to the fitness of the organism. They use evolutionary theory for explaining the adaptive value (also called function) and history (phylogenesis) of the behaviour. It leads to questions like: what are the fitness consequences of behaviour and what evolutionary process explains why it was selected? Two reminders are important for understanding the distinction between proximate and ultimate explanations. First, these two (causal) explanations complement each other but are different types of explanation which cannot be conflated. For example, how organisms cooperate is perhaps understandable in terms of 'brain mechanisms' or 'motivations', but these explanations at the proximate level do not explain why cooperation exists at the ultimate level, for they do not describe processes that clarify why cooperation is stable in a population. Thus if we notice that only humans reciprocate (a form of cooperative behaviour that is presumably absent in the other animals), then we have described the presence of a proximate mechanism in the human population. Second, there is a connection between ultimate and proximate explanations, namely that we can predict that the *process* of evolution will lead to a certain *product*, because phenotypes are the effects of genes. Evolutionary theorists emphasize that Hamilton's theory explains how the logic or rationale of inclusive fitness theory is enforced: the theory predicts that natural selection leads to organisms that are inclusive fitness maximizing agents (Grafen, 2006; 2008). Thus the product of evolution is an organism that behaves in such a way that it maximizes its direct fitness (the impact of behaviour upon personal reproductive success) and indirect fitness (impact upon the reproductive success of genetically similar individuals). This is the reason why evolutionary theorists use teleological language to describe the effects of genes. Some effects influence the probability that a gene will be replicated, either through effects in its carrier or in kin of the carrier. Genes with a positive effect will be perpetuated; alternative genes with less positive effects will be eliminated. Hence natural selection leads to effects which are in this sense beneficial for the gene and are for this reason called *functions* (see Haig, 2012). Others have argued that the effects of genes give the impression that they are *designed* (Gardner, 2009; West, El Mouden and Gardner, 2011). However, these suggestions may lead to misunderstandings because the differences between teleological and evolutionary explanations are not discussed by these authors. I shall discuss teleology and its place within the framework of inclusive fitness theory in Chapter 4.

1.6   The aim of this book

Before I elaborate the topic of this book, it may be helpful to set out first what this book does not attempt to do. It is not a contribution to the mathematical development of inclusive fitness theory. There are many excellent papers and books describing this theory. In Chapter 3, I shall summarize the essential insights necessary for understanding the problems discussed in this book.

The aim of this book is to extend the conceptual framework of inclusive fitness theory so that it is capable of explaining the social evolution of human nature. It is an extension, for the changes that I will discuss and elaborate do not alter the principles of inclusive fitness theory. In terms of the distinction between ultimate and proximate causation: the changes that I will advance concern proximate mechanisms. Inclusive fitness theory suggests that proximate mechanisms are selected because they maximize inclusive fitness. The sixth major evolutionary transition leading to human societies occurred because of language evolution. Hence the challenge is to explain, in terms of inclusive fitness theory, how and why linguistic behaviour evolved as a *new* form of behaviour. I shall argue that investigating how linguistic behaviour evolved out of animal behaviour is essential for understanding why humans became a unique creature among the apes. Hence by investigating how linguistic behaviours evolved, we will obtain a better understanding of human nature. I shall discuss the (possible) genetic adaptations during the early stages of language evolution and how they paved the road to the evolution of *Homo loquens*.

The development of an extended framework of inclusive fitness theory suitable for studying the social evolution of human nature requires also an analysis of conceptual problems. I have already mentioned the reason why: the essential difference between humans and other animals is the use of language. We are above all language-using animals; *Homo loquens* and only therefore *Homo sapiens*. The ability to use a language is the mark of having a mind, for it is because of language that we are rational animals, are able to reason and can think, feel and act for reasons. We would not possess the distinctive powers of the intellect and will if we were not able to use a language. Consequently, if we want to understand human social evolution and the impact of language on the evolution of the human mind, then we will discover how the mind and, hence, a self-conscious, culture-creating creature evolved subsequent upon language evolution. Such an inquiry also involves resolving conceptual problems, for it requires an answer to the question of what is meant here by the human mind. My answer, discussed in Chapters 2 and 5, is that the human mind that evolved is roughly understandable in terms of the Aristotelian conception of the mind. I shall also explain why it is for conceptual reasons mistaken to invoke the Cartesian conception of the mind. In order to elaborate this thesis, I shall contrast throughout this book the Aristotelian and Cartesian conception. I mention an example in order to clarify what is at issue here. The Aristotelian conception emphasizes that language enabled humans to form intentions, i.e. to act for reasons. Only humans can pursue goals and explain the reasons why they want to pursue goals. This ability to form intentions clarifies why humans evolved into rational creatures. In the Cartesian conception, by contrast, teleology (and forming intentions) does not play any role, for Descartes eliminated the concept of a goal. He argued that consciousness is the mark of the mind distinguishing humans from other animals. Hence to answer the problem how the human mind evolved is according to the Cartesian conception to answer the question how consciousness evolved.

These differences between the Cartesian and Aristotelian conception return in two readings of Darwin's critique of the argument from design. Some evolutionary biologists have argued that Darwin replaced teleology by natural selection. They believe that we can use Darwin's theory to explain why organisms are apparently designed, for natural selection has according to them design-creating capacity. According to an alternative reading of Darwin's critique of the argument from design, which I shall advance in this book, Darwin did not replace teleology by natural selection, but eliminated the concept of design and returned to the Aristotelian idea that goal-directedness is an intrinsic part of nature. Hence Darwin made in his work the first steps to integrating the Aristotelian conception and evolutionary theory. For Darwin, in contrast to many of his successors, saw clearly that evolution enables us to return to Aristotelian teleology. Yet Darwin did not pay much attention to the role of language evolution. Hence investigating the problem of how language evolved and what its implications are for the evolution of human nature will help us to integrate the Aristotelian conception of the mind and inclusive fitness theory.

Applying inclusive fitness theory to phenomena requires a clear conceptual understanding of the phenomena in which we are interested. Research into the ontogenesis and evolution of behaviour requires a correct conceptual framework characterizing the concepts of the phenomena investigated. There are many cases in which a conceptual contribution is not necessary. Yet there are also examples where conceptual problems cause confusion and obstruct scientific progress. Some of these conceptual problems are discussed and resolved in this book. I shall explain in Chapter 2 how conceptual problems can be resolved through analysis, i.e. through investigating the rules for use of words.

The conceptual foundation of human nature

2.1 Introduction

Humans are special because they are self-conscious, culture-creating beings. Yet how can we understand these unique features? How did they evolve after humans split with chimpanzees and bonobos from a common ancestor and what genes are involved in their ontogenesis? Many scientists assume that, as soon as we obtain more information about the genes that differentiate us from the other apes, we will understand how the human mind evolved. However, answers are not easy to get, for we have to rely on indirect methods. We cannot observe now the behavioural changes that were brought about by unique human genetic variations that arose thousands or millions of years ago.

There is another reason why clear answers to these questions are hard to get. Scientists and philosophers have argued that these questions involve not only empirical problems, but also conceptual problems. For asking how typical human features like self-consciousness evolved also invites the question of what is meant here by self-consciousness. When we study the evolution of, say, bipedal locomotion or the number of sweat glands in humans, it is clear what phenomena we are investigating and what is meant here with an emerging feature. Yet what is meant by saying that only humans are self-conscious beings? Is this a behavioural feature, or a property of the brain or mind? What methods can we use for investigating mental phenomena? The problem noted by many investigators is that it is unclear how we can measure mental phenomena. As the evolutionary biologist George Williams (1985, p. 22) put it: 'The power of positive thinking has never been measured in calories per second, nor a burden of grief in grams.'

Conceptual problems also concern (the evolution of) social behaviour. A comparison of the social behaviour of insects and humans clarifies why. In insect societies (such as the bee hive) there is a division of labour: most individuals are programmed to give up their reproductive functions and sacrifice themselves for others, while the queen takes care of reproduction. This separation between altruistic and reproductive functions is, for example, visible in the behaviour of the female bees of the royal court: their task is to care for the queen, reducing the chance that she will die of infectious diseases. This is comparable to the division of labour present in the human body: the gonads are for reproduction, other organs for the maintenance of the organism. The division of labour in both human bodies and insect societies is explicable in terms of inclusive fitness theory (altruism). Yet if we study human societies, there is no comparable division of labour in this sense (although one can argue that menopause, the phenomenon that middle-aged women become sterile, is part of an evolved reproductive division of labour in humans; see Foster and Ratnieks, 2005). Humans are not programmed for one function and all individuals are able to reproduce and to behave altruistically. In humans flexible abilities evolved that enable individuals to behave selfishly or altruistically depending on the circumstances. This raises the problem of how we can understand the origin of human behavioural flexibility, but it raises also the conceptual problem of what is meant here by 'behavioural flexibility'. Is there an essential difference between humans and bees? Are humans unique in this respect? Are humans unique because they, in contrast to other animals, have a mind (as Descartes argued) and are therefore able to display behavioural flexibility, i.e. are able to act at will? How, then, can we understand the evolution of the will? These questions show again that we are confronted with the conceptual problem of how we should conceive of animals, humans and their minds.

I shall argue in this chapter that, for understanding conceptual issues, it is useful to distinguish two conceptions of human nature, namely the Cartesian and Aristotelian conception. I shall also argue that only the Aristotelian conception enables us to explain the evolutionary origin of the gap between humans and the other animals. The advantages of the Aristotelian conception will be discussed at length in the chapters to come. In this chapter, I shall discuss some essential differences between the Cartesian and Aristotelian conceptions, and how these conceptions are used for explaining human and animal behaviour. I shall also explain what the aim of conceptual investigations is and how they are used for resolving philosophical problems.

2.2   Body and mind

In the modern era Cartesian dualism has provided the main conceptual framework of thought about humans, their minds and their bodies. It is also used for discussing the essential differences between humans and other animals.

Descartes was a key figure in the scientific revolution. He developed the so-called new philosophy of mind and matter (see Hacker, 1999; 2007; Kenny, 1968). The reason why Descartes developed an alternative to older conceptions was the new physics, the science of matter in motion. He distinguished two substances, body and mind, and thought that bodies are fully understandable in terms of the laws of the new physics. He also believed that biology was reducible to physics and treated animals, with the exception of humans, as machines: the essential functions of animal life can, according to Descartes, be understood in purely mechanistic terms. Humans are, according to Descartes, exceptional because only they have a mind. Hence human beings are in the Cartesian framework conceived as material bodies conjoined to mental substances (minds). Descartes argued that there are relations of two-way causation between the body and mind. The essence of the human mind is thought and thinking is the essential activity of the mind. Thought was defined by Descartes in terms of consciousness. He argued that thoughts are always transparent, i.e. humans are always aware of the thoughts they have.

Descartes' conception was revolutionary for two reasons. First, because he argued that thoughts are objects of *private* knowledge: if someone has a certain thought, then he apprehends that he has this thought in his mind (i.e. is aware of his thought). Other people, by contrast, cannot be aware of what goes on in someone's mind. They can only observe bodily movements, i.e. the externalities of essentially private thoughts. Hence they have to infer with the help of what nowadays is called a Theory of Mind (see, among others, Baron-Cohen, 1995; 2003), or by using an argument from analogy, what thoughts are occurring in the mind of the person whose bodily movements they observe. Second, Descartes' conception was revolutionary because of the way Descartes understood *thought*. He extended the traditional conception of thought: thought includes, according to Descartes, everything which we are aware of as happening within us. Whereas thought was traditionally defined in terms of the rational powers of the intellect and the will (see below), Descartes included in his conception of thought also sensation felt (e.g. feeling a pain), feeling emotions and desires and sensory awareness (see Descartes, 1985b, p. 195). The reason was that he assimilated thought with self-consciousness (we cannot think without being conscious of our thinking). Thus being aware of, for example, a pain was also, according to Descartes, an example of self-consciousness, for we cannot have a pain without experiencing that we have it. Consequently when we feel a pain (or feel emotions, imagine), then we have an experience and therefore always know that we have this experience. This conception of conscious experience led to the misconception of introspection as inner perception (see further section 2.7).

The Cartesian conception of animate nature and the human mind is accepted by many biologists and psychologists. It is used for understanding the evolution of the mind and raises what Crick and Koch (2002, p. 11) named 'the most mysterious aspect' of the mind/body problem: how did consciousness evolve? How can we understand consciousness as a property of a neural network? There is, of course, an important difference between Descartes' conception and the modern, materialist conception (see Bennett and Hacker, 2003; 2008): Descartes' mind/body dualism is rejected by materialists. Yet the essential point to notice is that Descartes' mind/body dualism is not replaced by (Aristotelian) monism, but is transformed into a brain/body dualism (the materialist conception is for this reason also called the *crypto-Cartesian conception*; see further Chapter 5). While Descartes' idea that the mind is an immaterial substance is rejected, the basic structure of the Cartesian framework is retained, for the mind is now identified with the brain. In the brain/body conception, it is argued that mental states are neural states of the brain. Note that, in contrast to the original mind/brain dualism of Descartes, there is no difficulty in the brain/body conception in envisaging causal interactions between the brain and the body, for both are now conceived as material substances. Hence the Cartesian difficulty (how can the material body interact with an immaterial mind?) appears to be resolved. Yet the problem is that the large part of the general structure of the Cartesian picture survives intact (see also Smit, 2012). For example, just as in the original Cartesian conception, it is believed that the relation of neural (mental) events to physical movements is causal; that the relation of impact of energy or matter on nerve endings causes changes in the brain (mind); that mental events and states (studied as neural states) are concealed; that action is movement caused by neural (mental) events; that mental predicates can be attributed to the brain (mind), etc. This has far-reaching consequences. For if we adopt the general structure of the original Cartesian framework, then we are inclined to reformulate the old Cartesian problems in terms of the modern vocabulary of neuroscience and (evolutionary) genetics. Problems that are in the original Cartesian framework conceptual problems appear then in the new framework as empirical problems. However, because the structure of the Cartesian conception is retained, they are still conceptual problems but now *disguised as empirical problems*. This statement charges me, of course, with the task of explaining in this book why these problems are conceptual ones which can only be resolved through conceptual investigations, not by empirical studies.

Aristotle was the first philosopher who broke with the dualist tradition in Greek philosophy (i.e. the dualism of Plato). He did not discuss body and mind (what Aristotle called *psuchē*) as two different substances, entities or parts of the organism. In Aristotelian monism body and *psuchē* are one (the *psuchē* is not a distinct entity but the form of the body). Aristotle made a distinction between things and living creatures with the principle that only living beings have a *psuchē*. It is important to note that the *psuchē* is, in contrast to the Cartesian conception of the mind, a *biological principle* and that the possession of *psuchē* is not a characteristic of human beings alone (see further Chapter 5). Yet Aristotle also differentiated humans from the other animals: only humans have a rational *psuchē* (the rational powers of the intellect and will). According to Aristotelians, this essential difference between humans and other animals is explicable in terms of language evolution: only humans use a language and have therefore the rational powers of the intellect and the will (Hacker, 2007; 2013). We are *Homo loquens* and therefore *Homo sapiens*, i.e. the wise ape. Whereas Descartes believed that thought or consciousness is the mark of the mind, neo-Aristotelians argue that *mastery of a language* is the mark of the rational *psuchē*. Hence they have to explain how and why linguistic behaviour evolved and why understanding language evolution is essential for understanding the rational powers of the human mind. This will be discussed in Chapters 5 and 7.

How does the constitutive role of language explain that other animals lack the typical human, rational powers? Animals have a *psuchē*, but their *psuchē* is considered relatively 'primitive' because animals are not language-using creatures. The horizon of animals' thought is determined by what animals can express in their non-verbal behaviour, whereas the horizon of human thought is determined by what humans can express in non-verbal *and* linguistic behaviour. For example, humans can think of specific dated events because they use a tensed language; animals cannot. Humans, and not the other animals, can reflect on the laws of nature, can fantasize about how things might have been and can imagine countless possibilities. And because humans, as language-using creatures, have the power of imagination, only they create works of art. As Karl Marx put it in *Das Kapital* (1867, vol. 1, chapter 7, section 1): 'A spider conducts operations that resemble those of a weaver, and a bee puts to shame many an architect in the construction of her cells. But what distinguishes the worst architect from the best of bees is this, that the architect raises his structure in imagination before he constructs it in reality.'

The limits of thought are (only) in humans set by the linguistic expressions of thought (see Hacker, 2001, chapter 2; Hacker 2007). Since language evolved from simple to more complex linguistic behaviour, we can expect that these limits evolved too. What humans could do with words expanded during the course of evolution resulting in a gap between humans and other animals. For example, humans started to think of the past and the future subsequent to using a tensed language (i.e. the use of tenses and of temporal referring expressions). They could then remember not only *where*, as other animals also can (exhibited in their seeking, homing and food-storing behaviour), but also *when*. As a corollary, whereas other animals can *prepare* for the future (e.g. they bury food for later consumption), humans can *plan* for the future (and can develop projects). Language evolution affected also the way humans can cope with errors they make. Animals have cognitive abilities and know therefore many things. They can think things to be so-and-so and later recognize that they were wrong and *rectify* their error. But they cannot reconsider and *correct* their belief, for this ability requires the use of a language (humans, as language-using creatures, can understand propositional truths).

This brief discussion of the (crypto-) Cartesian and (neo-) Aristotelian conceptions shows why it matters which conception we adopt for investigating the evolution of human nature. The materialist or crypto-Cartesian conception raises the problem of how we can explain the evolutionary origin of thought or consciousness as a distinct brain process in only the human species. This culminates in studying the neural correlates of consciousness and in attempts to find genes coding for brain processes involved in thought (i.e. concepts, ideas). It also results in attempts to explain, in terms of inclusive fitness theory, how the Theory of Mind evolved, enabling humans to predict the private mental states of others (for a discussion, see Smit, 2010a). According to the neo-Aristotelian conception, by contrast, we have to study language evolution and how linguistic behaviour evolved as an extension of animal behaviour, for language enables humans to do things with words and led to new forms of communicative behaviour. Humans have nowadays an innate ability to acquire a language, yet this second-order ability evolved and this raises the problem of how the ability arose as the result of genetic variations affecting brain and behavioural development.

Many scientists use the Cartesian and a few (the author of this book is one of them) use the Aristotelian conception when studying the evolution of the mind. We can evaluate their investigations in two ways. The first way is conceptual analysis: through investigating the use of words, terms and concepts, we can determine whether Cartesians and Aristotelians use concepts coherently. This is an important task, for empirical investigations aiming at discovering empirical truths presuppose the correct use of linguistic expressions. Only if the use of these expressions makes sense can we discover empirical truths (we cannot determine whether a stone or bacterium feels a pain, for the use of the concept of pain does not make sense here). The second is an empirical way: since we can use the Cartesian and Aristotelian conceptions for framing hypotheses about the evolution of the animal and human mind, we expect that only the conceptually coherent framework is capable of generating testable empirical predictions and of explaining new facts. I shall briefly discuss an example.

Humans with chimpanzees split from our common ancestor 6 million to 7 million years ago. Modern techniques enable us to study genetic differences between humans and chimpanzees (see Somel, Liu and Khaitovich, 2013). We can also use the fossil record for studying possible effects of unique human genetic variations. It is estimated that there are about 15 million to 20 million genetic changes that are specific to humans, but most of them probably have no functional impact. An example that is known to have an impact are the different duplicates of the gene SRGAP2 (Charrier *et al.*, 2012; Dennis *et al.*, 2012). It was found that these duplicates are involved in brain maturation: they lead in the human species to a higher density of spines on dendritic cells in the human cortex. One of the duplicates, that evolved 2 million to 3 million years ago, reduces the activity of the original gene. Because the original SRGAP2 regulates brain maturation in chimpanzees, and since the new duplicate inhibits the activity of the original gene, the new duplicate contributes to an extended brain development. For it takes now longer for the human brain to mature (this is called neoteny). This may explain why humans, compared to chimpanzees, have a longer juvenile period and have better learning skills. It is interesting to note that this duplicate evolved at a time when the hominids started to use stone tools. Hence it is possible that the duplicate evolved because it enhanced inclusive fitness as the result of improved learning skills.

How should we conceive of these duplicates and their effects in terms of the (crypto-) Cartesian and (neo-) Aristotelian conceptions? If we assume that the human variants of SRGAP2 are involved in the evolution of the human mind, then Cartesians can argue that the gene is involved in the emergence of human thought or consciousness. Their task is then to demonstrate how the duplicates, affecting brain maturation, lead to brain processes correlating with conscious experiences. Aristotelians can argue that the duplications of SRGAP2 have been constitutive of typical human learning skills and, hence, extended behavioural flexibility of the early hominids. One can imagine that the duplicates enabled children, because of the better learning skills, to learn the function of stone tools because they could rely on the pre-existing knowledge about the stone tools of their parents (see further Chapter 7). Inclusive fitness theory explains why the duplicates evolved: they enhanced inclusive fitness because the enhanced learning skills resulted in new forms of communicative behaviour (e.g. teaching and learning skills). Because the duplicates enabled humans to engage in new forms of communicative behaviour, Aristotelians can argue that they contributed to the gradual emergence of the gap between humans and other animals (and may have been involved in the early stages leading to language evolution). Aristotelians, in contrast to Cartesians, emphasize that SRGAP2 is not directly involved in the evolution of thought and self-consciousness, for thought and self-consciousness presuppose complex linguistic skills that evolved during the later stages of language evolution. The reason why I prefer the Aristotelian hypothesis is not only because I believe that the available empirical data confirm only this hypothesis (and because the hypothesis raises interesting questions), but also because I believe that the Cartesian hypothesis is not testable at all. Cartesians have it wrong for conceptual reasons.

2.3   Conceptual investigations

Conceptual investigations are useful when scientists encounter what they call semantic problems. I have already mentioned two reasons why they are confronted with conceptual problems. First, there are different conceptual frameworks, in particular the Cartesian and Aristotelian frameworks, which are not always clearly distinguished in discussions. Mixing these frameworks up creates conceptual problems. Second, conceptual problems arise because conceptual propositions sometimes have the misleading form of empirical propositions. This creates confusion if conceptual problems are treated as empirical problems: we sometimes mistakenly try to solve conceptual problems through empirical studies even though there is no empirical problem to solve. I shall discuss some examples later in this section.

   Conceptual analysis can be used to disentangle empirical and conceptual propositions so that we get a clearer picture of what the conceptual and empirical issues are. And if we have resolved conceptual problems, then we remove incoherence and are able to frame better hypotheses. Resolving conceptual problems requires clarity about whether they concern technical or ordinary terms. We can therefore differentiate two different conceptual contributions.

First, the contributions made by several evolutionary theorists clarifying how concepts should be used in the context of inclusive fitness theory. I have already discussed some examples in Chapter 1, but will mention here another. Trivers (1971) was the first theorist who distinguished two forms of cooperation: between related and unrelated individuals. He called cooperation between unrelated individuals 'reciprocal altruism'. Based on the discussions in Chapter 1, the reader will directly remark that this is a misnomer, for altruism only occurs when individuals are related. 'Reciprocity' has therefore been suggested as a better term. It is captured in the principle: 'You scratch my back, and I'll scratch yours.' Trivers' distinction raised the empirical problem whether cooperation between non-kin in animal societies can be understood in terms of reciprocity. At first glance, many investigators thought so, but research over the last thirty years shows that reciprocity is absent or rare in animal societies (Clutton-Brock, 2009). When animals cooperate, both partners gain immediate benefits and there is no evidence that cooperation has a temporary fitness cost (as required if it was an example of reciprocity). Hence it is concluded that cooperation in animal societies is an example of mutualism (both partners benefit without a fitness cost) or manipulation. This brief discussion of reciprocity shows what is meant by conceptual clarity about technical terms and why it improves empirical studies. The absence of reciprocity in the other animals raises the conceptual problem why it occurs mainly in humans. I have already discussed the answer: reciprocity requires the use of a tensed language (see further Chapters 5, 7 and 8).

Second, the contribution made by investigators who clarify the rules for the use of the concepts involved in descriptions of social behaviour and cognition. When we investigate human behaviour, we use ordinary concepts such as 'emotion', 'sensation', 'memory', 'thinking', 'feeling' and 'willing', and these are not technical or theoretical terms. Clarification of these concepts is a prerequisite for posing fruitful questions about the social evolution of human nature.

Conceptual investigations are useful because they clarify what belongs to the conceptual net and what to the empirical fish that we catch with it. Conceptual propositions express how concepts are used and they structure a framework (or network) through internal relations. Without these internal relations empirical hypotheses that are framed within the confines of the framework would disjoint. Conceptual sentences provide therefore coherence: they establish internal connections between the major structural features or elements of our conceptual framework (see Strawson, 1966). Yet how do we investigate the structure of a conceptual framework

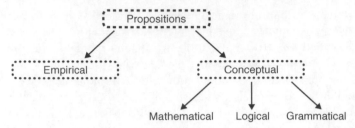

Figure 2.1: A subdivision of the different types of propositions.

and its relation to the external world? In order to answer this question, it is useful to discuss first the differences between empirical and conceptual propositions, and the subdivision of conceptual propositions into mathematical, logical and grammatical propositions (see Figure 2.1).

Empirical proposition are, for the most part, *bipolar*, that is they can be true and can be false. Whether a proposition is true or false depends on what the case is. For example, the proposition 'This rose is red' is true if the rose is, in fact, red. Hence understanding an empirical proposition is to know that if the proposition is true, then things are thus-and-so, and also to know that if it is false, that things are then not so. By contrast, a conceptual proposition is not bipolar and does not have the possibility of being true and the possibility of being false. For example, the proposition 5+5=10 cannot be false. To know a mathematical proposition is not to know that things are thus-and-so, but to know a rule, i.e. a rule for the transformation of empirical propositions concerning quantities or magnitudes. If a child understands that 5+5=10, then it knows that if one has two bags with five marbles each, then one has ten marbles in all. And if the child knows that both bags contain five marbles, then it knows, when it counts them up and finds only nine, that it has made a mistake or that one marble has vanished. Grammatical (also called: conceptual) propositions also lack the possibility of being true and the possibility of being false. For example, the proposition 'Every rod has a length' is true but cannot be false, for there is no such thing as a rod without a length (Wittgenstein, 2009 [1953], sections 251–253). Similarly, 'Every human being has a body' cannot be false, for there is no such thing as a human being without a body. Hence one cannot, for conceptual reasons, investigate empirically whether a rod has a length or whether someone has a body. If someone says that he has found a rod without a length or says that he has discovered a human being without a body, then he transgresses the bounds of sense and utters nonsense.

Grammatical propositions have the potential to mislead us since they have the form of empirical propositions. The proposition 'Mary has a body' has the form of the empirical proposition 'Mary has a book.' But whereas she may possess a book (we can check whether Mary has, in fact, a book), she does not possess a body, for Mary does not own a body and, hence, it makes no sense to check whether Mary possesses a body (Mary does not lose her body whereas she can lose one of her possessions). Hence the proposition 'Mary has a body' is not an empirical but a conceptual truth (and presupposes the rule or grammatical proposition 'A human has a body'). What, then, is meant by the phrase 'having a body'? By saying that we have a body, we often mean that we have certain somatic qualities (see Cook, 1969; Hacker, 2007, chapter 9; Kenny, 1988b). We are beautiful, athletic or emaciated as the result of disease. Thus the proposition 'A human being has a body' is a grammatical proposition, but we can use it in formulating empirical propositions (such as 'Mary has an athletic body') which are capable of being true or false.

It follows that conceptual problems are not comparable to and should not be treated as empirical problems. And because they are unlike empirical problems, they cannot be solved with the help of the methods from science. They do not call for new discoveries but require conceptual clarity. And if conceptual clarity is provided, then the problems will be resolved or dissolved. This is done through investigating the rules for the use of words. Another way of expressing this is saying that (some) philosophical problems arise because the sentences used for formulating these problems express misconception. These sentences transgress the bounds of sense, i.e. they express nonsense. When we *realize* that these sentences are nonsense, then we also realize that the problem posed is ill-conceived.

In order to avoid possible misunderstandings, I discuss two elucidations. First, it is sometimes thought that a description of the rules for the use of words is a form of linguistics. This is a misconception, for resolving conceptual problems through investigating the rules for the use of words is not like describing the rules of grammar, for example, that a transitive verb must be followed by an expression in the accusative case. Conceptual problems are not resolved by pointing out grammatical mistakes, but by showing that the bounds of sense are transgressed. For example, 'What is the relation between mind and brain?' is grammatically correct, but the question posed cannot be answered because the rules for the use of the word 'mind' are transgressed. For if we investigate the use of the word 'mind', we find that the mind is not an entity that can stand in a relation to the brain. Second, it is misguidedly thought that resolving conceptual problems invokes finding the truths expressed in ordinary language, i.e. the common sense expressed by the man in the street. But the rules investigated through conceptual studies are not studied to highlight 'empirical beliefs', for the conceptual problems are not empirical problems but conceptual ones that arise as the result of misunderstanding or misusing the rules for the use of words. These words may be words used in our everyday discourse, but also the technical words used in science.

2.4   Resolving problems

I discuss three examples in order to show how conceptual problems are resolved through conceptual investigations. I have taken these problems (and have added some material) from Schroeder (2006), who discusses many others.

2.4.1   Identity and property

The British philosopher Bradley (1893) was puzzled by the problem that properties of a substance do not really determine what it is. For example, a lump of sugar (the subject) is white, hard and sweet (three predicates we ascribe to the subject). But sugar is, according to Bradley, not all that, for a thing is not any one of its qualities: sugar is not identical to white, sweet or hard. Thus, it appears that for such a statement to be true, the predicates have to be the same as the subject (i.e. 'Sugar is sugar'). The problem noted by Bradley is that this statement is uninformative. Thus there appears to be a dilemma: if you predicate what is different (e.g. that sugar is white), you say that the subject is what it is not, and if you predicate what it is (e.g. 'Sugar is sugar'), you say nothing at all. What went wrong here? The paradox can be resolved through noting that the word 'is' is used here in two different ways. It is used as 'is the same as' (George Orwell is Eric Blair) and is used to ascribe a property to the lump of sugar. Bradley's problem originates in his misguided assumption that the word 'is' must always indicate identity.

2.4.2   Definitions and facts

There is one thing in our metrical system that is used as a sample: the standard metre in Paris. It does not have a special property, but has a special role in our system. And because it is used as a sample, we can deny that the standard metre is one metre long, for we do not use measurement to find out what its length is. But we can also say that it is one metre long, for that is its length! That sounds paradoxical, but this paradox dissolves as soon as we realize that there are two uses of the word 'is' involved here. 'The standard metre is as a matter of fact one metre long', and 'The standard metre is by definition one metre long.' If we deny that the standard metre is or is not one metre long, then we use the word 'is' in its factual sense. This sounds paradoxical, because there is also the analytic sense (the definition), in which it is correct to say that the standard metre is one metre long. Thus the alleged paradox is only a way of drawing attention to the distinction between definitions and empirical statements.

A similar problem may arise if we conflate 'This is red' as an empirical proposition (which we can paraphrase as 'This object is as a matter of fact red') and 'This is red' as an ostensive definition. In that case the proposition is used as an explanation for the use of the word 'red'. It can be paraphrased as, 'This *colour* is red.' Suppose that someone sees a tomato, points to it, and says 'This colour is red.' Then he explains the meaning of red. The tomato is then used as a *sample*, i.e. a standard of comparison, which may be used to determine whether other objects in the surroundings are red too. For they are correctly said to be red if they are the colour pointed at.

Wittgenstein has remarked that an ostensive definition may be used as a substitution rule and is therefore comparable to a (linguistic) translation-rule that explains the meaning of a word in terms of another (Wittgenstein, 1958, p. 109). For example, we can explain the word 'bachelor' as an 'unmarried man'. An ostensive definition explains the meaning of a word in a similar way. For instead of saying 'this object is red', we can say 'This object ☞ is that colour.' The deictic gesture and the sample (e.g. the tomato) are then used as a symbol interchangeable with the word 'red'. Hence the substitution rule teaches us here that 'This object ☞ is that □ colour' = df. 'This object is red', just as 'bachelor' = df. 'unmarried man'. But it is important to keep in mind that there are differences between ostensive and analytic definitions of words. For example, we can always say that a 'bachelor' is an 'unmarried man', but in order to explain that 'red' is 'this colour' (together with a deictic gesture and a sample), the sample must be available (see further Baker and Hacker, 2005, Essay 5; Hacker, 2001, chapter 9; Smit, 1989, chapter 3).

2.4.3 *Micro- and macro-features*

Scientific discoveries are a source of philosophical puzzlement about the meanings of words. For if the meanings of words change as the result of scientific discoveries, are we then still talking about the same thing? For example, chemists discovered that water is H_2O. Can we say that the meaning of water is now determined by 'the real nature or essence' discovered by chemists and that the older definition was wrong, i.e. water is a transparent, tasteless, drinkable liquid that occurs in rivers, rain and lakes? And should we replace our everyday definition by the scientific one? A way of resolving these problems is through making a distinction between definitions in terms of the micro- and the macro-features of water. For this clarifies why the scientific definition does not replace the everyday definition. Two fictive examples illustrate why this distinction is helpful. Suppose that chemists discover a substance that is indistinguishable from water except that it has a different molecular structure; would we still call it 'water'? Because it has a different micro-feature, we will probably say that it is not water. But because the substance has similar macro-features, we will call it a new type of water, since the substance has the same use in most practices. Now suppose that chemists discover that a black, viscous and smelly substance has the molecular structure H_2O; would we then say that the substance is identical with water? We would probably not, for we cannot use this substance for the same purposes for which we use water.

These two examples clarify why there are not many reasons for preferring only one definition in all contexts. Chemists often prefer the definition at the micro-level, whereas we use the definition at the macro-level in our everyday life. Thus, as the result of scientific discoveries we know more about all kinds of substances, but this only shows that we are able to describe the same substances by increasingly sophisticated concepts. That scientific discoveries do not make older definitions and explanations superfluous, also applies to definitions and explanations used in biology, medicine and psychology. Suppose that a patient complains of something that causes pain in his chest, and that the doctor diagnoses that the pain is caused by tissue necrosis in the left ventricle. Then it would be absurd to argue that the doctor's explanation is at cross-purpose with the explanation of the patient and that we should only use the scientific one.

2.5   The (crypto-) Cartesian conception is incoherent

I said in section 2.2 that Cartesians have it wrong for conceptual reasons. I can now clarify why: Cartesians use expressions that transgress the bounds of sense. Consider the Cartesian assumption that the immaterial mind interacts with the material body. This is not an empirical statement that can be tested, since it is not intelligible to talk about an immaterial substance. A substance is something which we can identify, but there are no criteria of identity of the mind defined as an immaterial substance. We do not know how to identify the mind as an immaterial substance, how to measure this substance, and so forth. And since there are no criteria to identify the mind, it is senseless to say that the mind has causal powers and can, therefore, interfere with physical processes. What kind of empirical evidence can someone provide if he argues, for example, that the mind causes a voluntary movement? How can we determine that an immaterial substance causes this movement if we cannot identify this substance?

Modern materialists (i.e. crypto-Cartesians) replace Descartes' mind/body dualism by a brain/body dualism. They argue that the mental states of the mind can be studied as neural states (see, e.g. Crick, 1995). For instance, when explaining how children acquire empirical knowledge, they argue that the brain 'receives' information, 'processes information' and uses the processed information for 'planning movements and actions'. To know something can according to crypto-Cartesians be investigated as a state of the brain. However this conception of the brain as an information processing organ is, just as the original Cartesian conception, incoherent. I discuss three reasons (see Hacker, 1995; 2006). First, it is a misconception to assume that the senses receive 'unprocessed information'. Light waves impinging on our retinae and sound waves agitating our eardrums are not 'unprocessed information', since they are not information at all (when someone tells me that *p*, or when I read that *p*, then I acquire information, but the stream of photons and sound waves are not information in that sense). Second, it is misguidedly assumed that the brain is an organ that 'processes information' (see also Chapter 6). Of course, we cannot see unless the visual cortex is functioning normally, but we see with our eyes, not with our brain. Likewise we cannot remember something unless our hippocampus is functioning normally, but we remember something, not our hippocampus. And we cannot walk unless the motor cortex of the brain is functioning normally, but that does not mean that we walk with our brain and that the brain is the organ for locomotion, as the legs are. Neither the brain nor its parts are organs for exercising psychological powers in the sense in which eyes are organs for seeing or ears are organs for hearing. Suppose that I see a red tomato, do I then experience a red tomato in my brain? Is this experience a neural state of my brain? Saying so is incoherent, for there is no such thing as experiencing a red tomato *in* my brain. It does not make sense to answer the question where I experience the red tomato by saying: 'Here', while pointing to my head (as opposed to pointing at the fruit in the garden). Similarly, it cannot be said that the hippocampus is the locus of remembering, for an answer to the question 'Where and when did you remember that . . .?' is given by saying: 'While I was in the library'; not by saying: 'In my hippocampus; where else?' Third, to know something is mistakenly assumed to be in a neural state. Perception is a primary source of knowledge: when children perceive that something is so-and-so, they acquire information of how things are. But this is not to be in a certain neural state, for to know something is ability-like, and hence more akin to a potentiality than to an actuality (a state). When someone knows something, then he is able to do a wide range of things: he can inform others, answer questions, can correct others, find, locate, identify and explain things and so forth. To forget that something is so-and-so is not to cease to be in some state, but to cease to be able to do certain things. Of course, the activities of neurons are a causal condition for acquiring and remembering a piece of information, for without the activities we could not acquire and remember knowledge. But if someone wants to determine whether I know a great deal of the theory of evolution, then he will not study the neural state of my brain, but whether I am able to answer an indeterminate array of questions about the subject matter. These answers are criteria for saying that I understand the theory and that I am able to solve problems, can correct errors, can tell to others appropriate facts, etc. The conclusion may be that I know some parts in detail and others only reasonably well. But whatever the results, it will not be concluded that I believe (as opposed to know) things in detail, for to say so would be incoherent (see further Hanfling, 2000, chapter 6).

Neo-Aristotelians argue that the normal functioning of the senses and the brain is a causal condition for acquiring empirical knowledge. Mutations affecting, for example, our discriminatory abilities affect therefore the possibility that someone can acquire empirical knowledge. For instance, the colour blind cannot use all of our colour words because they cannot make the distinctions we make, as the result of a mutation of a (chromosome X-linked) gene. Hence they lack the ability to use some of the samples and rules we use for explaining colour words, cannot correct and explain mistakes in the use of some colour words and cannot therefore determine the truth of some empirical propositions. What would happen if only the colour blind were to populate the planet? A part of our colour language would then disintegrate, for the distinctions we make could no longer be used because the colour blind cannot distinguish them as we do.

2.6 Receiving versus acquiring information

Neo-Aristotelians emphasize that, instead of saying that the senses of children *receive* information, it is better to say that children *acquire* empirical knowledge by using their senses. The organs of sight are exercised in a large variety of ways which we discriminate in our vocabulary (see Hacker, 1991): we see, spot, glimpse, peep, glance, peek, catch sight of; we look for, watch, look at, stare at, gaze, observe; we examine, scrutinize, inspect, scan, survey; and we discern, make out, descry, recognize, etc. To argue that we can investigate all of these different perceptual capacities in terms of 'receiving and processing information' is a misconception. It is far more interesting to investigate the differences between these capacities and to ask when the need to distinguish them did arise. Some of these are activities that take time, for example, inspecting a room, examining an artefact or scanning a scene. These activities can be interrupted or completed. Others are not always activities that are completed, for example, when we watch, look or stare at things. And there are examples that are not activities but more akin to achievements (see Ryle, 1949, chapters 5 and 7), for instance, when we discern or spot something. I shall investigate the evolutionary origin of some words involved in empirical knowledge in Chapters 6 and 7, and discuss here briefly an example to clarify what is at issue.

Suppose that a mother wants to teach her child the properties of a toy, then she will ask the child to focus his attention on her activities and the toy. Since language has evolved, she can use expressions such as 'Watch carefully!' or 'Examine what you can do with this toy!' Hence she orders or invites the child to engage in an activity. Note that the parent cannot use here the expression 'See this toy carefully!', for seeing is itself not an activity. Hence if the child understands that he is invited to join an activity (aiming at conveying information), then he will not summarize what he is going to do by saying: 'I am going to see . . .', but by saying: 'I am going to examine . . ..' Hence in a situation in which parents want to convey knowledge requiring the attention of their children, we expect gestures and perceptual words to evolve enabling tutors to instruct children (i.e. enabling them to explain that children should use their senses and cognitive faculty).

Note again that there is a profound difference here between the (neo-) Aristotelian and (crypto-) Cartesian conception. Because Aristotelians argue that mastery of a language is the mark of the mind, answers to the question of how children acquire empirical knowledge are, in part, answering the question of how they learn to use perceptual predicates (Hacker, 2012). Aristotelians argue that learning the use of perceptual predicates is preceded by learning to use words, norms of representation and descriptions. Hence mastery of the perceptual vocabulary (the use of verbs of perception and their cognates) presupposes antecedent mastery of an observational vocabulary of perceptibilia, i.e. that children already know how to name things and their visible properties and how to describe objects. The perceptual vocabulary is being mastered when children learn to ask and to respond to questions such as: 'Can you see . . .?' or 'Did you hear . . . ?' And in response to assertions of how things perceptibly are, they learn to answer the question 'How do you know?' by replying that they saw something. For example, if someone asks a child: 'How do know that Mummy is in the garden?', he can answer that he saw her there. Children learn then to use verbs of perception as operators on descriptions of perceptibilia. At the same time, they learn the use of these verbs in the contexts of commands, interrogations and questions, i.e. they learn to order ('Look!', 'Listen!', 'Daddy smell!'), to ask or query ('Can you see?', 'Did you hear?'). When they have mastered this expansion of the perceptual vocabulary, it becomes possible to understand error and illusion, i.e. they learn that perceptual descriptions are not always right. Observation conditions are sometimes suboptimal, the sense organs are sometimes defective (temporarily or permanently) and the objects of perception are sometimes deceptive and look or sound other than they are. When they make errors, their parents will correct these and will explain how children can correct errors: by looking again or by improving the observation conditions (moving closer, turning on the light). Children learn then the use of the operators 'It seems to me as if', 'It appears as . . .', 'I think it is a . . .', and so forth. They begin to understand that these operators are used to qualify observational sentences and are used when the employment of the senses and the cognitive faculty was not optimal. Notice that this (Aristotelian) picture of how children acquire empirical knowledge is the opposite of what Descartes thought. Descartes argued in the second *Meditation on the First Philosophy* that he could doubt his senses, but not that things *seem* to be so-and-so.

This brief discussion of how children acquire empirical knowledge shows that conceptual investigations teach us that there is a sequence in how the perceptual vocabulary expands. It starts with the use of gestures, norms of representation and words, and ends with the ability to use all kinds of operators qualifying observational sentences. It follows that the neo-Aristotelian ('acquiring empirical knowledge by using the senses') and the crypto-Cartesian conception ('receiving and processing information in the brain') have repercussions for how we should study the ontogenesis and evolution of empirical knowledge. While crypto-Cartesians ask how the brain constructs knowledge (investigated as neural states of the brain) subsequent to processing information received by the sense organs, Aristotelians emphasize that humans acquired the ability to use perceptual predicates subsequent to the acquisition of the ability to name things and to use norms of representation by using their senses. I shall argue in Chapter 7 that the ability to communicate about perceptibilia evolved during the course of evolution as an extension of the use of gestures such as pointing, for these gestures facilitated the use of norms of representation.

The differences between the neo-Aristotelian and crypto-Cartesian conception of empirical knowledge have all kinds of ramifications in other areas. I mention here one example. Neo-Aristotelians emphasize that children are from early on experimenters. As self-moving creatures capable of acting at will (see further Chapter 5), they can, by using their hands, pick up things, look at objects from different angles, push or pull objects, drop them or throw them away, and so forth. Hence in opposition to the crypto-Cartesian receiver (the brain as an organ that processes information), we have the neo-Aristotelian agent or experimenter (Hacker, 2005b), for handling and manipulating go together with perceiving. This difference between the two conceptions also clarifies why neo-Aristotelians argue that children learn to use the concept of a cause as experimenters. First they acquire the notion of (what philosophers call) substance causality; later they acquire the notion of event causality. The essential difference between substance and event causation is that substances are, whereas events are not, space-occupants (Hacker, 2007, chapter 3). Children can therefore see the occurrence of substance causality (e.g. a stone breaking a window or a cat killing a mouse), but have to infer event causality. And because children are experimenters, they are 'causers': they not only see a stone break a window, but can cause the change themselves by throwing a stone. Because the verb 'cause' in event 'E1 causes event E2' does not stand for anything observable, event causation is less easy to understand.

2.7 Feeling pain

Conceptual investigations are indispensable for scientific progress. The reason is that many words used by evolutionary theorists are not scientific terms but words which we use in ordinary discourse. In contrast to terms such as 'gene' and 'natural selection', these words have a meaning independent of how science defines terms. Examples are words such as 'body', 'mind', 'sensation', 'disease' and 'health'. Clarifying the rules for the use of these words helps us to develop a coherent framework characterizing the concepts of the phenomena investigated. I shall discuss first the example of the sensation of pain and then health and disease (sections 2.7 and 2.8). Through these discussions I shall elaborate the differences between the Cartesian and Aristotelian conceptions. I shall also explain why neo-Aristotelians believe that the Cartesian conception is incoherent. In section 2.9 I return to a major theme in this book: how does the Aristotelian conception differentiate human from animal social behaviour?

How do we know that we have pain? And how do we know that other persons have a pain? The Cartesian answer to the first question is that we know that we have pain because we feel it. For Cartesians assume that we experience pain in our mind conceived as a separate, immaterial substance. Yet the problem is that feeling or perceiving pain is not something done with one of our senses. It is tempting to say that we perceive pain with an inner sense, i.e. through introspection. But introspection is not a form of observing, for there are no inner sense organs. Hence we cannot ask someone to have a closer look at his pain if he says that he has pain. Moreover, we cannot improve the conditions so that he is able to perceive the pain better, as we can in the case of observing something in the outer world. Hence there are obvious differences between observing an object in the outside world and feeling pain. They raise the question of why we think that feeling pain is based on a perception. A conceptual analysis of what is meant by 'feeling pain' may help us here (see Hacker, 2005a).

When we feel a penny in our pocket, the penny is felt by touch. Since 'I feel a pain in my foot' has the same form as 'I feel a penny in my pocket', there is the temptation to believe that feeling pain is done with an inner sense. But pain is not felt with an inner sense. What, then, is meant by the utterance 'I feel a pain'? The answer to this question is surprisingly simple. When we say that we feel a pain, then we mean that we *have* a pain. The utterance 'I feel a pain' is a learnt extension of natural pain behaviour, i.e. the expression of pain in our behaviour. We have learnt the expression 'I feel a pain' as, in part, a replacement of natural expressions, like crying, screaming out of pain, or saying 'au'. Of course, we feel pain in our body, for sensations have a location in the body. Suppose that I stub my toe against something and it hurts. Then I have pain in my toe and feel this pain. But if someone observes the toe that I am clutching, then he does not see pain (neither do I). He just observes a swollen, red toe. Nevertheless it is a painful toe and that is what I *express* in non-verbal and verbal behaviour: 'I feel a pain in my toe.' This expression of my sensation is, of course, dependent on nerves sending action potentials from the toe to my brain. Without these action potentials I would not feel any pain. But when the action potentials enter my brain, then I do not somehow and somewhere perceive pain in my brain or my mind with an inner sense (as Cartesians assume), for there is no such thing as perceiving a sensation in the brain. They result in an expression in my (non-) verbal behaviour and this expression is a criterion for my being in pain.

The differences between feeling pain and perceiving an object in the outer world have consequences for how we should study phenomena if the mind is involved. First, it follows that the distinction between 'what happens in the mind' when someone has pain or is ill and what is expressed in non-verbal and verbal behaviour, should not be conceptualized as a distinction between a hidden inner and an observable outer. For the inner is expressed in the outer, as the analysis of the sensation of pain demonstrates. The pain I feel (and my weariness as the result of a disease) is expressed in my behaviour and, hence, is visible for others. There is no external, causal relation between inner and outer phenomena (as Cartesians assume), but an internal relation. Behavioural manifestations of the inner, i.e. non-verbal and verbal behaviours, are the criteria for applying psychological predicates to someone and for saying that he has an inner life. When someone screams with pain or says that he has pain, or when he is fatigued and says that he is ill, then these manifestations of his sensations are for us reasons for saying that he has pain or is sick. Of course, someone may be lying or be insincere (e.g. a malingerer who complains of pains but never goes to a medical doctor). His expressions of sensations are then not reliable manifestations. But the point to notice is that the untruth of what he is saying is not comparable to the untruth of someone who is mistaken when he reports about an empirical fact. Someone may make a mistake when he observes something, but when I am lying I do not make a mistake while observing my inner life, for there is no such thing as observing my inner life. Second, it follows that someone who has pain does not have privileged access to his pain in the sense that he alone has knowledge about his pain while we, as outsiders, can only indirectly know that he has pain. For thinking that only the subject of pain can know that he has pain misguidedly presupposes the Cartesian distinction between hidden mental phenomena accessible only to the subject and public, observable bodily movements. It leads to the mistaken thought that we, as outsiders, can only acquire knowledge about the pain of the other by either a theoretical inference (we infer that a person is in pain and use observations of bare bodily movements to test our hypothesis), or by an analogy with our own experience of pain.

We can conclude that the conceptual confusion created by Descartes resulted from his identification of the mind with consciousness, his association of consciousness with the private, and his conception of the private as a domain to which the subject has privileged access. These errors have far-reaching consequences and some of them will be discussed in later chapters (and have been discussed by others; see among others Candlish, 1995; Hacker, 1987; 1991; 2002; 2007; 2013; Hyman, 1989; 1991; Kenny, 1989; Rundle, 1997; Smit, 2010c). I mention here one consequence. If we adopt the neo-Aristotelian instead of the Cartesian conception, then we can clarify in what sense animals are conscious too. Animals are conscious for they undergo periods of sleep or unconsciousness (and awake and regain consciousness); because their attention is caught by objects and events in their environment; and because they enjoy or suffer states of consciousness such as feeling contentment, hunger and pain. Yet neo-Aristotelians also argue that there is an essential difference between animals and humans, for only a language-using creature is self-conscious. It does not mean, as Descartes mistakenly believed, that only humans have private access to their feelings and thoughts, but means that humans can reflect on what they are doing or undergoing, on their reasons for acting, thinking or feeling, on their motives and motivations, on their likes and dislikes and on their character traits and relations with others. Because they can reflect on these, they are able to understand them better and can sometimes modify them in the light of their evaluations. Hence the use of a language enables humans to reflect and to weigh their deliberations and this is partly constitutive of being self-conscious creatures.

2.8   Health and disease

The Cartesian and Aristotelian conceptions of pain return in discussions of the concepts of health and disease. According to the Aristotelian conception, we use behavioural manifestations to determine whether someone is ill, for someone who is sick (and realizes that he is ill) expresses his illness in non-verbal and verbal behaviour (e.g. by saying that he does not feel well). Just as in the case of pain, he does not perceive his illness because there is no inner sense enabling him to observe his disease. 'I feel sick' is, just as 'I feel pain', an *expression* of being sick. When someone is ill, then he tells us that things are not going with him as they normally do, that he feels warm or cold, that he has pain and is fatigued, that he has vomited and feels nausea, and so forth. Activities like walking, breathing, focusing one's attention on something, which are normally done without much effort, now require attention and energy. Hence he has problems with doing the things that he could do before he felt ill.

When someone is ill, then there is something wrong with his body. His body malfunctions, constrains his normal activities and limits his opportunities for action. As a result of these constraints he gets worried and this may be a reason to consult a medical doctor. If there is a treatment for the disease, then he is subjected to this treatment (aimed at curing the disease) and has to meet the prescription of the medical doctor.

Malfunctioning organs are causes of diseases (see also Chapter 4). There are several causes of why organs malfunction. For example, malfunctioning lungs may be caused by bacteria or viruses (with the result that someone is short of breath), or by having been a heavy smoker for a long time, or by having been a coal miner. Medical researchers investigate in these and other cases what the precise mechanisms are that lead to malfunctioning lungs. Note that in the Aristotelian conception the concepts of a goal and function are essential to the understanding and characterization of an organ: it is said that malfunctioning organs may cause disease and constrain individuals in developing and executing goal-directed activities.

In the Cartesian conception the concept of a goal is not used for understanding health and disease. The reason is that Descartes objected to the use of this concept and argued that explanations in science should be purely mechanistic. In Descartes' mechanical world picture the human body is conceived as a machine. Disease is, according to this conception, explicable as a defective machine. Yet how do we know that a part of the machine is 'malfunctioning'? Aristotelians use here the concept of a purpose (a normal functioning organ is partially constitutive of the health of the organism), but this possibility is ruled out in the Cartesian conception. Some Cartesians are inclined to refer here to the design of the organism: disease occurs if there is a deviation from the species design. Just as we can use the design of the machine as it is, for example, drawn on paper for determining why a machine is 'malfunctioning', Cartesians argue that we can use the species design to determine whether someone is sick. This conception is, however, problematic, for the concept of design does not belong to Darwin's theory but to natural theology (or creationism), a tradition that preceded evolutionary theory. According to natural theology, the great Designer created the various species according to a plan or design. If we knew this plan, then we would understand the different designs of species and, hence, could determine deviations from the design (resulting in diseases). But we make then certain assumptions which are, according to Darwin's theory, incorrect.

First, the concept of design presupposes that species are perfectly adapted to their environment (according to the plan of the great Designer). Yet organisms are only optimally adapted to their environment. If, for instance, the environmental conditions change, we may be less well adapted to the environment. For example, we have an immune system that protects us against many pathogens in our environment, but this system is not capable of coping with HIV causing Aids. There is now selection going on resulting in an improved immune system, but the result will not be a perfect system because there is always the possibility that new variations evolve among pathogens. Second, variations in a population occurring at a low frequency are not simply deviations of a species design (and must therefore be called diseases). There are neutral variations within populations which do not affect fitness; there are so-called genetic polymorphisms (variations co-existing in the population) resulting in various morphological, physiological and behavioural forms. There are different forms within one individual depending on the circumstances (individuals develop a tanned skin in the summer and a white one during the winter; they have more red blood cells when they are living in the mountains). There is also variation within the life cycle of an individual, for example, the binding capacity of the foetal haemoglobin differs from the adult form; within each individual a unique HLA type is generated through somatic recombination and mutation affecting the chances of survival of the individual. All the examples show that it is difficult to distinguish health and disease, normal and abnormal, with the help of the concept of a species design.

Another Cartesian argument faces similar problems: some Cartesians argue that textbook physiology points in the direction of a species design. For in the textbooks there is often only a discussion of one design: the working of a species-specific organ is discussed. But do these authors discuss an ideal type that is subsequently used by medical doctors to differentiate health and disease? The answer is no. For these authors are supposed to provide information about the working of an organ *and* the function it has. That there is variation is, given the purpose of the textbook, at first not relevant for explaining the function of an organ. Is it important to discuss that a certain individual has, compared to another, three extra muscle cells in his left ventricle? That is not the case as long as these extra cells do not affect the action of the heart. But if someone has fewer cells in his ventricle (as the result of apoptosis) and if tissue necrosis affects then the heart's functioning (its pump function), then there are good reasons for mentioning and studying this case when someone's health is affected. However, note that this is an Aristotelian insight: a malfunctioning organ is then the reason for studying the working and morphology of the heart. Hence if Cartesians defend this interpretation, then they advocate the Aristotelian conception of health and disease.

The Cartesian conception of health and disease is also at variance with the concept of Darwinian fitness. When evolutionary biologists talk about differences in the fitness of individuals, then they do not mean differences in health but differences in the number of offspring (see further in Chapter 3). Because certain characteristics of individuals affect numbers of offspring, characteristics that increase these numbers will be selected. If this fitness criterion were to be used for defining health, then this would lead to absurd consequences: someone who has fewer children would be less healthy than someone who has more. Is someone who decides not to have children ill? Does someone's health depend on his partner (for children are produced with an opposite sex-individual)? Answers to such questions show why the Cartesian conception results in conceptual incoherence.

The Aristotelian framework clarifies here why Darwinian fitness and health are two different concepts. It tells us that we should not conflate the evolutionary origin of a characteristic and its function or goal. For example, we can investigate the evolutionary origin (and maintenance, modification) of the heart in terms of its contribution to fitness. But an answer to the question if and how a heart contributes to the health and well-being of an organism requires teleological explanations. For a well-functioning heart is partially constitutive to the health of the organism (of course, a healthy organism may have higher fitness because it produces more offspring). Hence neo-Aristotelians do not deny or ignore teleology, as Cartesians do, but differentiate teleological from evolutionary explanations.

2.9   Needs, desires and the will

Only the neo-Aristotelian conception is coherent and useful for understanding the social evolution of human nature. The essential insight is that language evolution was constitutive of the evolution of the rational powers. Language evolution has expanded therefore the range of situations in which cooperation could evolve and resulted in extreme forms of division of labour characterizing human societies. The key factor here is to understand how mastering the use of a language enables children to extend and enlarge natural behaviours also displayed by other animals. I shall discuss the needs, desires, wants and the will to illustrate how we can use the neo-Aristotelian conception for understanding differences and similarities between human and animal behaviour.

Animals, in contrast to plants, use their senses and are self-moving creatures. They can pursue the objects of their wants. Whereas plants only have *needs* (e.g. they need water), animals have *wants* and exercise two-way powers or abilities, i.e. they can act but can also refrain from acting as they please. Yet although animals have wants and can pursue the objects of their desires (water, food, sex), the horizon of their wants is limited since the objects of their desires are constrained by their limited mental capacities: they can choose and have preferences, but cannot deliberate. For animals do not use a language, cannot give reasons for their preferences and cannot reflect on these reasons. Hence when we say that animals make a decision, we mean that they simply terminate a state of indecision. Since they do not use a tensed language as humans do, they cannot explain and justify their decisions (and actions consequent upon decisions) by forward and backward-looking reasons. By contrast, humans can give reasons and can do something because it is desirable or obligatory given certain reasons or values. Only a language-using creature can reason and deliberate, weigh the pros and cons of facts that it knows in the light of its desires, goals and values.

Animals are not language-using creatures and lack therefore the ability to form intentions: i.e. they cannot give reasons for what they are doing or going to do and cannot, therefore, be said to be accountable for their actions. And if they cannot form intentions, they cannot act altruistically or egoistically either. For example, it does not make sense to say that a male mouse is an egoist when he kills the young (infanticide) of an unknown female, because it is not an example of an intentional act. Infanticide reduces the fitness of male competitors and increases the chance that the female comes in oestrus again. Interestingly, males do not kill the young of the female with whom he has copulated and who stays in his proximity, because his 'killing behaviour' is then inhibited for two months as the result of a neuroendocrine mechanism (it takes two months before his offspring begin a life independent of nursing). No less interestingly, the scent of an unknown male induces in female mice the termination of a pregnancy, reducing the costs of investing in a pregnancy that does not contribute to their fitness (because the young would be killed by the male after parturition). Hence instinctive behaviours are not manifestations of egoism, for they are not examples of behaviours that license us to attribute volition to animals.

What, then, is meant by 'egoism'? It means that behaviour is motivated by self-interest. Suppose that I donate a certain amount of money because I am motivated to alleviate suffering or because I am a wealthy person and wish to be applauded for my noble generosity. The latter donation is an example of vanity; the first example not. Although the two acts are the same, the two persons invoke different reasons for their acts: wishing to alleviate suffering versus wishing to be applauded. Note that this is simply the definition of 'wanting': 'I did it because I wanted to' may mean 'I did something voluntarily.' If asked, I may give reasons if I acted intentionally.

Some theorists hold that acting intentionally requires a causal explanation because it is an example of satisfying a sub- or unconscious desire. If I choose to do something, so the argument runs, then the reason I give masks the fact that I have a desire to do something, and acting is a way of satisfying this desire. But is 'doing something because one wants to' like satisfying a desire? If I want to do something, then I form an intention, i.e. I pursue a certain goal and can give reasons why I pursue that goal (for example, I donate money because I care about the poor and the sick). This is not like satisfying a desire. When I am thirsty, then drinking a glass of water satisfies my desire for water. Hence in this case there is a link between a bodily state (being thirsty), wanting something (water) and an action aimed at the satisfaction of the desire. The problem is that 'donating money because I care about the poor' is altogether unlike satisfying a bodily desire. Moreover, the idea that all human voluntary actions are motivated by unconscious desires is not supported by evidence. Of course, humans often act unconsciously, but that does not mean that we always act because of sub- or unconscious desires (see further Dilman, 1983; 1984).

2.10   Conclusion

I have discussed in this chapter why a discussion of the conceptual foundations of human nature is indispensable for understanding the social evolution of humans. Some problems concerning the evolution of human nature concern conceptual problems, although they are not always recognized as conceptual problems. I have explained the reasons why: conceptual problems appear in the crypto-Cartesian framework as empirical problems because Descartes' mind/body dualism is transformed in a brain/body dualism. There are two ways for resolving these problems: through invoking Aristotelian monism, for this helps us to see Cartesian misconceptions, and through using conceptual investigations. We can use conceptual investigations for disentangling empirical truths from conceptual truths concerning human nature. I have discussed some examples and have explained differences between empirical and conceptual propositions. It is important to remark here that many philosophers (misguidedly) believe that there is no distinction between conceptual (or analytical) and empirical propositions, but this will not be discussed in this book (this is discussed at length in, e.g. Hacker, 1996b and Schroeder, 2006).

Differentiating empirical from conceptual truths has as a consequence that we can disregard theories or hypotheses for two reasons. First, they may be false because of (an accumulating set of) data contradicting predictions and/or because they are not capable of explaining new-discovered facts. Well-known examples in life science are the group selection hypothesis of Wynne-Edwards, the blending theory of inheritance and the instruction theory in immunology. Second, they may not be testable at all because of the incoherent use of concepts. Well-known examples are the (Cartesian) idea that mind and brain interact, the idea that introspection is a form of inner observation and Wundt's innervationist and James' ideo-motor theory for explaining the development of voluntary movements (see Hacker, 1990b; 1996a). Conceptual investigations differentiate here sense and nonsense. Combinations are, of course, possible: I shall discuss in later chapters why the (Lamarckian) theory of acquired characteristics, the social intuitionist model of (moral) emotions, the nativist model of language and moral evolution are confronted with both conceptual and empirical problems. In others papers and books (see, e.g. Smit, 2010a; 2010c) I have argued that the conflict theory of autism and schizophrenia and the ToM-theory of social cognition are conceptually incoherent and not supported by empirical data.

There is a general reason why most theories are conceptually incoherent: they use a variant of the Cartesian conception of human nature. The Cartesian conception is not only incoherent because of its mind/body dualism (or brain/body dualism), but also because there is no space in this conception for teleological explanations. Descartes argued that we should study animate nature without the concept of a goal. Evolutionary theorists have later argued that Darwin replaced teleology by natural selection. I shall argue in Chapter 4 that they have misunderstood Darwin's critique of the argument from design. Darwin did not replace teleology by natural selection but showed that we can understand goal-directed phenomena with the help of natural selection. He rehabilitated the Aristotelian conception of goals. This misunderstanding of Darwin's critique is another reason why the Cartesian conception has dominated discussions in life science in the modern era.

CHAPTER 3

Inclusive fitness theory and genomic imprinting

3.1 Introduction

Darwin's contribution to our understanding of why such diverse living creatures populate the earth is well known. He showed that we do not need to invoke supernatural forces for understanding why the different creatures on earth evolved. Darwin demonstrated that the evolution of life is understandable as the product of a single force: natural selection. By using natural selection as an explanatory principle, we can understand the origin, maintenance and transformation of simple and complex forms of life. Of course, there are other processes such as neutral evolution as the result of genetic drift that contribute to evolution (see Kimura, 1983), but Darwinians believe that natural selection is the major driver of the complexity of life.

In this chapter I shall discuss the main extension of Darwin's theory: inclusive fitness theory. This extension is important for understanding the social nature of humans because Hamilton's theory is capable of explaining the evolution of cooperation. Inclusive fitness theory was developed by Hamilton as an extension of ideas advanced by the population geneticist Fisher. I shall start therefore with a brief discussion of how alleles are selected in populations. How do models from population genetics describe the dynamics of gene-frequency change under the action of natural selection? Next, I shall explain how Hamilton's theory extended Fisher's model so that it can be used for understanding the evolution of genes underlying social behaviour. The essential insight of the extension is that, over time, organisms will evolve to maximize their inclusive fitness. Hamilton (1964, p. 24) summarized this insight as follows: 'the situations in which a species discriminates in its social behaviour tend to evolve and multiply in such a way that the coefficients of relationships involved in each situation become more nearly determinate'. Applied to kin selection, it means that natural selection favours organisms that are able to subdivide other organisms as kin or non-kin and to adjust their behaviour accordingly.

Inclusive fitness theory has been used for explaining phenomena at different levels of organization. I shall discuss some examples and shall elaborate one particular example at length: genomic imprinting, i.e. the phenomenon that the expression of alleles in an organism depends on the sex of the parent. The sex-specific expression of imprinted genes is stage-specific: some imprinted genes are expressed during prenatal stages; others are expressed during neonatal, juvenile and later stages. The so-called kinship theory of genomic imprinting uses inclusive fitness theory to explain the effects of imprinted genes on child development. At first this theory was developed as an extension of Trivers' theory of parent–offspring conflict; later this theory was modified and extended so that it also encompasses inbreeding avoidance and dispersion. I shall elaborate this theory and shall also explain how it can be used for understanding disorders.

3.2   Inclusive fitness theory

Evolution is the process of adaptation occurring via the action of natural selection. Natural selection is the result of differential reproductive success of individual organisms. It operates on genes correlating with the characteristics that are associated with differences in reproductive success. Evolution presupposes therefore that there is genetic variation in a population. If there is no variation, there is no selection. Darwin's theory predicts an increase of those variations in the population that increase reproductive success. Variations that increase the reproductive success of their carrier will therefore get fixated in the population. Hence adaptations evolve as the result of the accumulation of heritable characteristics in a population that are correlated with reproductive success. Fisher (1930) showed that the selection of favourable genes results in an increase in the mean fitness of the population. Fisher's model explains why organisms, over successive generations, appear increasingly well adapted for achieving reproductive success.

Hamilton realized that organisms are not always selected to maximize their own reproductive success. He showed that genes correlating with characteristics can not only increase in a population through increasing the fitness of an individual, but also through increasing the fitness of other individuals carrying copies of those genes. For example, when an individual helps a close relative raising his or her children, then fitness benefits of the relative will contribute to the fitness of the helping individual. Natural selection will lead to organisms that are adapted to maximize their, what

Hamilton called, *inclusive fitness*, i.e. the sum of the personal fitness and the fitness that results from helping a relative.

For understanding inclusive fitness theory, it is useful to define some technical terms. The term *fitness* is defined as the numbers of offspring that an individual produces that survive to adulthood. An important point to notice here is that whether a behaviour is beneficial or costly is defined on the basis of (1) the lifetime fitness consequences of the behaviour (not just the short-term consequences) and on the basis of (2) the fitness of individuals relative to the whole population. The latter follows from the fact that the term is used in the context of population genetic investigations. I shall discuss an empirical example as an illustration.

Visser and Lessells (2001) studied the effects of clutch size on the fitness of the great tit. In experiments they manipulated the clutch size so that the birds had to raise two extra chicks. There were three experimental conditions: either (1) two extra nestlings were added to the nest, or (2) two extra eggs were added to the nest, or (3) the female was induced to lay two extra eggs (by removing the eggs of the clutch so that the female laid new eggs, and then adding the old eggs to the new ones). The effects of these manipulations were that in all conditions the tits raised two extra tits (i.e. four tits instead of two). The results of the experiments showed that the number of young who survived to breeding age did not differ between the three conditions. Hence if we were to calculate the fitness effects during one breeding season, there were no significant effects. But there was a difference in fitness effects when the next breeding season was taken into consideration. Visser and Lessells found that the manipulations affected female survival. Females in the third condition who experienced the extra costs of laying two extra eggs (besides the costs of having to incubate them and to raise the chicks) had the lowest survival rate. Hence if we calculate the lifetime fitness of the individuals, then we have a different result compared to calculating the reproductive success per brood. This illustrates what is meant by lifetime fitness consequences and why the lifetime reproductive success is used for calculating fitness.

Defining fitness on the basis of lifetime consequences implies that components of the fitness of an organism, for example, the health and survival of the organism, are not distinct parts of the individual's fitness. This does not mean that these are unimportant. On the contrary: when we study, for example, the stages of a life cycle of a particular species, it is useful to pay attention to these components (see further Chapter 4). Yet they are not part of the formal models of inclusive fitness theory, for at the population level, i.e. the fitness-maximising effects of genes

Table 3.1: A classification scheme for social behaviour, based on Hamilton (1964) and West, Griffin and Gardner (2007).

	Effect on recipient	
Effect on actor	+	−
+	Mutual benefit	Selfish
−	Altruism	Spite

relative to conceivable alternatives, the effects of genes are weighed only via their lifetime inclusive fitness effects. Hence defining fitness in this way is a simplification, but it is useful for it enables evolutionary theorists to develop theorems and to derive predictions from these theorems. It is important to recall the reason why: these models define social behaviour solely in terms of inclusive fitness effects. This is possible, for evolutionary theorists distinguish ultimate and proximate causal explanations (see Chapter 1). Their models discuss only ultimate explanations of social behaviour.

Hamilton distinguished four types of social behaviour. These are defined as behaviours that have fitness effects on both the individual that behaves (*actor*) and on another individual (*recipient*; see Table 3.1).

Selfishness is beneficial for the actor and costly for the recipient. It increases the fitness of the actor and its evolution can easily be understood in Darwinian terms. Selfishness is pervasive and visible in forms of aggression (for example, cannibalism, infanticide, territorial exclusion and exploitative interactions such as brood parasitism). *Mutual benefit* is beneficial for both the actor and the recipient and its evolution is therefore also easy to understand. It is, like selfishness, pervasive and underlies the evolution of the egalitarian transition (see Chapter 1), but is also manifest in, for example, cooperative foraging. *Altruism* is costly for the actor and beneficial for the recipient. It is manifest in the fraternal transition (somatic cells giving up their reproductive abilities and aiding the germ cells; sterile workers helping the queen in eusocial insects). The evolution of altruism is, however, less easy to understand in terms of Darwin's original theory because it is costly for the actor. Hamilton argued that altruism evolved because actor and recipient share genes: the act may indirectly contribute to the fitness of the actor because the act increases the fitness of the recipient and therefore indirectly the fitness of the actor (see further below). *Spite* requires for the same reasons a special explanation, for it is costly for both the actor and the recipient. For understanding why spite occurs, it is important to recall that the four forms of social behaviour are defined in terms of an abstract, mathematical model. The four behaviours are defined on the basis of their effects on lifetime fitness effects relative to the whole population. The latter requirement clarifies why spite may evolve: it can occur if the relatedness between two individuals is less than average in the population. West, Griffin and Gardner (2007) mention the release of costly toxins by bacteria in order to kill competing bacteria as an example.

3.3 Cooperation

Hamilton used these definitions for investigating the problem of why cooperation occurs, for they help us to answer the question why an organism should engage in cooperation if that activity benefits other individuals. Notice first why there is a problem to solve. Suppose that organisms engage in cooperative behaviour and assume that a free-rider (or cheater) enters this group. A free-rider is a non-cooperating individual who benefits from the cooperative behaviour of the other group members without paying any costs associated with being cooperative. For example, when individuals share food in the group, then a free-rider is an individual who benefits from this cooperative behaviour without contributing to food gathering: he or she does not pay the acquisition costs. Consequently, genes correlating with free-riding or cheating have greater fitness in this population than the genes correlating with cooperative behaviour. Hence we can predict that these free-riding genes will increase in frequency although this may lead to a decline in population fitness. Hence the problem that we face when studying the evolution of cooperation, is what mechanisms solved the problem of free-riders and cheaters.

Table 3.1 shows that there are two principles that explain the evolution of cooperation: mutual benefit and altruism. For understanding the difference between these two forms it is useful to subdivide fitness effects into two categories: *direct* fitness and *indirect* fitness effects (West, Griffin and Gardner, 2007). Indirect benefits are achieved if the cooperative behaviour is directed towards an individual who carries genes for cooperation (altruism). The best-known way this can occur is if cooperation is directed on kin (genealogical relatives). By benefiting a close relative, an organism is indirectly passing copies of its genes on to the next generation. Hence it increases indirect fitness. Direct benefits are obtained if both actor and recipient experience a fitness benefit, yet cooperation faces then the problem of free-riders. We can therefore predict that the evolution of cooperative behaviour requires extra mechanisms enforcing cooperation and suppressing cheaters. For example, enforcement may be the result of policing or punishment imposing costs on non-cooperative free-riders and cheaters. In the case of humans, *reciprocation* may be a mechanism involved in cooperation. Trivers (1971; see also Axelrod and Hamilton, 1981) showed that cooperation can be favoured in repeated reciprocal interactions, for this creates the possibility that individuals preferentially aid those who have helped them in the past. Cooperation is then selected if the short-term cost of being cooperative is outweighed by the long-term benefit of receiving cooperation. Hence cooperation as the result of reciprocal interactions is, within the framework of inclusive fitness theory, an example of a direct fitness benefit (explicable as mutually beneficial). There are two forms of proximate mechanisms distinguished by which cooperation is preferentially directed at cooperative individuals (see Nowak and Sigmund, 2005): *direct reciprocity* (help those that help you) and *indirect reciprocity* (help those that help others).

What proximate mechanisms are involved in altruism? Hamilton suggested two general mechanisms for explaining indirect fitness benefits: limited dispersal and kin discrimination. *Limited dispersal* (also called population viscosity) can generate high degrees of relatedness in a population because it keeps relatives together. Because it is likely that neighbours are relatives, inclusive fitness theory expects that cooperative behaviour evolves. An example is the social behaviour of the social amoeba *Dictyostelium discoideum* described in Chapter 1 (section 1.3). *Kin discrimination* is the other possibility: an individual can distinguish relatives from non-relatives and uses this ability to preferentially help relatives. It can occur through genetic and environmental cues. Genetic cues include, for example, the odour produced by the scent glands which has been shown to play a role in ants and certain mammals. Environmental cues are derived from a shared environment or prior association and probably involve learning.

Hamilton discussed one other mechanism of indirect fitness, namely what Dawkins (1976) later called *the green beard effect*. It is a form of indirect fitness benefit, but the difference between the green beard effect and other examples is that the cooperative behaviour is directed towards those who share the same cooperative gene. Assume that the gene coding for the cooperative behaviour causes a green beard in its bearer, making it easy for other carriers to direct cooperative behaviour only to other carriers. Hence the green beard effect is, in effect, an assortment mechanism: by means of a green beard individuals preferentially direct cooperation towards carriers of the same gene (or of a number of tightly linked genes). Although this mechanism can evolve, mathematical models have shown that green beards are rare (see Gardner and West, 2010). I mention here three reasons. First, cheaters that display the green beard but not the cooperative behaviour will be selected and invade and overrule the population, resulting in a non-cooperative population. Hence a population consisting of individuals displaying the green beard is susceptible to the invasion of mutants. Second, since the green beard gene is not related to the other genes in the genome, there is the potential for a conflict between the green beard gene and the other genes in the genome. Inclusive fitness theory expects strong selection to suppress the action of the green beard gene by the other genes. Third, the green beard effect is an example of a recognition mechanism and a pre-requisite for recognition is genetic variability. Since individuals with more common cues are more likely to be helped, the more common cues will be driven to fixation while rare cues will be eliminated from the population. Hence the genetic variability will disappear, killing the recognition system, explaining why recognition systems based on genetic cues are rarely found in nature. Organisms often use cues for kin recognition which have genetic variability because of other evolutionary forces. The best-known example is kin recognition based on the MHC (Major Histocompatibility Complex) system of vertebrates (called the Human Leukocyte Antigen or HLA-System in humans). Host–parasite co-evolution causes genetic variability and this variability is used by animals (including humans) for kin recognition.

An example showing that organisms are capable of adjusting their behaviour in response to kin discrimination based on MHC is discussed by Neff (2003). In the bluegill sunfish (*Leponis macrochirus*) competition between males has led to the evolution of two distinct strategies. Males termed 'parentals' defend nest sites, attract females and care for the eggs and the newly hatched offspring. The other strategy is termed 'cuckolders': they steal fertilizations from parentals either by darting into the nests at the critical moment of spawning, or by mimicking females. Investigations show that cuckolders can fertilize 80% of the eggs released by a female (they participate in 8% of the spawnings). However, Neff showed that parentals use cues for paternal certainty and adjust their caring behaviour accordingly. It is highly probable that parentals use MHC in olfactory discrimination of kin.

3.4 Hamilton's rule

Because indirect fitness often involves kin, this part of Hamilton's theory has been called kin selection theory. Whether an act enhances the inclusive fitness can be calculated with Hamilton's rule: $rb>c$ (see Box 3.1). In his rule

Box 3.1: Hamilton's rule

Hamilton's rule of kin selection theory can be derived using the Price equation. From the following equation (discussed in Box 1.1):

$$\Delta\overline{w} = \mathrm{cov}\left(\frac{w}{\overline{w}}, w\right) = \frac{\mathrm{var}(w)}{\overline{w}} \qquad (3.1.1)$$

one can show that the direction of selection acting upon a character of interest is given by the least-squares regression (slope) of relative fitness on the genetic value of the character $\left(\beta_{\frac{w}{\overline{w}},g}\right)$.

Hamilton noted that for understanding the (inclusive) fitness of an individual, we also have to take the genes of the individual's social partners into consideration. Fitness may be affected by genes in the focal individual (g) and by genes of the partners (g'). The least-squares regression can be partitioned so as to describe the partial effects of both sets of genes:

$$\beta_{\frac{w}{\overline{w}},g} = \beta_{\frac{w}{\overline{w}},g.g'} + \beta_{\frac{w}{\overline{w}},g'}g = \beta_{g'g} \qquad (3.1.2)$$

Because of the effects of the genes in social partners, the effect of the genes in the focal individual is $\beta_{\frac{w}{\overline{w}},g.g'} = -c$. This is the personal cost of the social behaviour. And because of these effects, the partners benefit as recipients from the effect of the focal individual's genes: $\beta_{\frac{w}{\overline{w}},g'.g} = b$ (the benefit of being a recipient of social behaviour). Because the genetic association between the social partners is $\beta_{g'g} = r$ (the kin selection coefficient of genetic relatedness), the condition for the behaviour to be favoured $\left(\beta_{\frac{w}{\overline{w}},g} > 0\right)$ yields Hamilton's rule: $rb-c > 0$.

There are two approaches to kin selection, the so-called neighbour-modulated fitness approach to kin selection, where b is interpreted as the benefit of receiving help *from* social partners upon the reproductive success of the focal individual ($\beta_{wg'.g}$), and the inclusive fitness approach, where b is interpreted as the benefit *to* social partners ($\beta_{wg.g'}$).

It has been shown that the two approaches always yield the same result.

FURTHER READING

Frank, S. A. (1998) *Foundations of social evolution*. Princeton University Press.

Gardner, A. (2008) The Price equation. *Current Biology* 18: R198–R202.

McElreath, R. and Boyd, R. (2007) *Mathematical models of social evolution: a guide for the perplexed*. University of Chicago Press.

Wenseleers, T., Gardner, A. and Foster, K. R. (2010) Social evolution theory: a review of methods and approaches, in T. Székely, A. J. Moore and J. Komdeur (eds.), *Social behaviour: genes, ecology and evolution*. Cambridge University Press, 132–158.

c is the fitness cost of the actor, b is the fitness benefit to the recipient and r
is their genetic relatedness. The c in this equation captures the direct fitness
effects and the rb captures the indirect fitness effects. Hence cooperative
behaviour evolves if the benefits to the recipient, weighed by the genetic
relatedness of the recipient to the actor, outweigh the costs to the actor.

The value of r is often calculated with the help of simple Mendelian
genetics. For example, r=½ in the case of full brothers, and r=¼ in the
case of half brothers. It is important to keep in mind that these calculations
are helpful approximations but are from an abstract theoretical point of
view incorrect. The reason is mentioned above: Hamilton's rule is a
mathematically derived rule at the population level. Hence one should
not say that r is 'the chance that a recipient has a certain allele, given that
the actor has it', for there is the possibility that the recipient and the actor
have the same allele because the allele is common in the population. It is
therefore better to say that r is the increased chance that recipient and actor
have the same allele relative to the population mean (see Gardner, West
and Wild, 2011). Only when the allele is rare in the population can we use
the simple Mendelian calculations. Another way to highlight the same
point is saying that it is genetic similarity that matters according to
inclusive fitness theory, rather than kinship. This also clarifies why green
beard altruism is part of inclusive fitness theory, but it is not an example of
kin selection.

Inclusive fitness theory has been successfully used to explain phenomena in several areas of life science. Examples are sex allocation, kin
discrimination, parasite virulence, inbreeding (avoidance) and dispersion,
selfish genetic elements, alarm calls and parent–offspring conflict. The
theory also explains cooperation within groups. Understanding how
inclusive fitness theory is used for explaining group selection is important, for it offers an alternative for the older group-selection hypothesis of
Wynne-Edwards (1962) and others. I shall briefly spell out the difference
between the old and new theory. The older theory stated simply that
group selection occurred because cooperative behaviour was beneficial
for the whole group. The problem with this hypothesis is that such
groups are susceptible to the invasion of a non-cooperative cheater (see
above). The new theory, by contrast, shows that cooperative groups
evolve if the problem of free-riders is solved. To put it in a general
statement: if between-group selection for cooperative behaviour exceeds
within-group selection against non-cooperators, then cooperation is
selected. The two possible mechanisms of the new theory of group
selection are discussed above: direct and indirect fitness benefits. For

example, group selection as the result of indirect benefits occurs if the organisms constituting the group are genetically identical (the fraternal transition). Hence investigating group selection is just an application of inclusive fitness theory. Yet it is important to keep in mind that when a higher-level collective evolves, the forces at lower levels have to be very weak (Gardner and Grafen, 2009). Recall that theory and empirical data show that selection within a collective is a strong force creating non-cooperators (see Box 3.2).

Box 3.2: Multilevel evolution

Multilevel selection is about selection within and between groups. This approach to social evolution begins by considering that the entities in the parent and offspring populations are social groups rather than individual organisms. For example, an entity i may be part of an individual but also part of the group to which the individual belongs. We shall use therefore two subscripts, i and j, to refer to the jth group and the ith individual within group j respectively. Thus the ith entity in the subpopulation j is labelled ij, and w_{ij} denotes the number of copies that are made of it in a given unit of time.

In the context of population genetics, changes in allele or gene frequency can affect the average fitness of a group but also of an individual. If we use the Price equation (discussed in Box 1.1), then the change in the average gene frequency (g) can be written as a function of the mean fitness and mean gene frequency in the jth group as:

$$\overline{w}\Delta\overline{g} = \text{cov}(w_j, g_j) + E_j(w_j\Delta g_j) \qquad (3.2.1)$$

This equation now describes selection on the groups in the population. But multilevel selection also assumes that there is selection within each group. For understanding the extension of the Price equation to multilevel selection, it is important to note that the term on the left-hand side ($\overline{w}\Delta\overline{g}$) looks like the second term on the right-hand side ($w_j\Delta g_j$). On the left-hand side we have the average fitness in the population times the change in frequency of the allele or gene in the population; on the right-hand side we have the average fitness in group j times the change in frequency of the allele or gene in group j. Hamilton noted that we can expand the term $E_j(w_j\Delta g_j)$ to capture the effects of within-group selection, because:

$$(w_j\Delta g_j) = \text{cov}(w_{ij}, g_{ij}) + E(w_{ij}\Delta g_{ij}) \qquad (3.2.2)$$

The point to notice is that the right-hand side of the equation is again a version of the Price equation, but now one level lower in the selective hierarchy, for it describes within-group selection. If we substitute this equation in the previous one (3.2.1), we get:

$$\overline{w}\Delta\overline{g} = \text{cov}(w_j, g_j) + E(\text{cov}(w_{ij}, g_{ij}) + E(w_{ij}\Delta g_{ij})) \qquad (3.2.3)$$

The first covariance term captures the effects of the gene on group success; the second covariance term captures the effect of the gene on the relative success of individuals within a group. The term $E(w_{ij}\Delta g_{ij})$ in this equation accounts for deviations due to other processes than selection. If this term is zero (there is no mutation, etc.), and since $\text{cov}(w, g) = \text{var}(w)\,\beta(w, g)$, then a gene for a social trait is selected for if:

$$\overline{w}\Delta\overline{g} = \text{var}(g_j)\beta(w_j, g_j) + E(\text{var}(g_{ij})\beta(w_{ij} \cdot g_{ij})) > 0 \qquad (3.2.4)$$

The right-hand side has partitioned the selection on the character into between-group selection (first term) and within-group selection (second term) components. The equation shows that the change in the character is neither wholly determined by selection within groups nor by selection between groups. Multilevel selection is a mixture of both. In some situations cooperators may experience a within-group disadvantage against cheaters; in other situations selection between groups may be strong enough to overpower the effects of within-group selection.

FURTHER READING

Hamilton, W. D. (1975) Innate social aptitudes of man: an approach from evolutionary genetics, in R. Fox (ed.), *Biosocial anthropology*. London: Malaby Press, 133–155.

McElreath, R. and Boyd, R. (2007) *Mathematical models of social evolution: a guide for the perplexed*. University of Chicago Press.

Okasha, S. (2006) *Evolution and the levels of selection*. Oxford University Press.

Wenseleers, T., Gardner, A. and Foster, K. R. (2010) Social evolution theory: a review of methods and approaches, in T. Székely, A. J. Moore and J. Komdeur (eds.), *Social behaviour: genes, ecology and evolution*. Cambridge University Press, 132–158.

3.5 Resolving conflict

Within the framework of inclusive fitness theory, there is the potential for conflict whenever r<1. Notice that, even when r=1 (for example, the cooperative behaviour of cells in a multicellular organism arising out of a fertilized egg cell), there are still opportunities for conflict, for it is possible that as the result of mutations genetic differences arise. Hence the mutation rate and the number of cell divisions necessary to 'produce' the adult organism are in these cases the essential factors predicting the occurrence of conflict. A well-known example is cancer in large organisms like humans. At the level of the cell, a cancerous cell out-competes other cells because of the enhanced cell proliferation. The cell has therefore an evolutionary advantage, but at the level of the individual (the group) there will be selection against cancerous cells since they reduce the fitness of the whole organism. Inclusive fitness theory predicts the suppression of cancerous cells since they reduce inclusive fitness.

The fact that cancer occurs at a relatively high frequency in the human population seems to contradict these predictions. Note, however, that most cancers are age related (they occur at old age) and also note that more humans get older nowadays. These age-related cancers hardly affect fitness because they occur at a time that most individuals would have died in natural conditions because of extrinsic mortality (e.g. infectious diseases). Disposable soma theory explains why the relatively high frequency of human cancers does not contradict the predictions of inclusive fitness theory (see Kirkwood and Austad, 2000).

I have explained in Chapter 1 that there are two mechanisms for resolving the potential for conflict. These are *self-limitation* and *coercion*. The difference between them is linked to the distinction between fraternal and egalitarian transitions and, hence, to altruism and mutualism. Self-limitation occurs in fraternal groups, i.e. when members are related. Kin selection theory predicts then that selfishness will be held in check because of the costs imposed by selfishness. The point to notice again is that self-limitation requires relatedness: only if lower-level units are related, is there the opportunity for self-limitation. Coercion is the other form of holding selfishness in check and there are several mechanisms: for example, policing, punishment and dominance. It occurs in egalitarian groups, but also in fraternal groups if the mutation rate is high (or if r<1, e.g. in insect societies). Notice that coercion cannot evolve in egalitarian groups as a form of altruistic behaviour, since the lower-level units are not related.

A well-known example of coercion in egalitarian groups are the mechanisms involved in the selective inheritance of mitochondria (Hoekstra, 1990). Conflict occurs because the genes of mitochondria and the host cell are unrelated. Recall that mitochondria were originally free-living bacteria and later became cell organelles in a host cell. Assume sexual reproduction and biparental inheritance of mitochondria, i.e. they are transmitted to the next generation by both the egg and sperm cell. In this case there is no mechanism that ensures equal transmission of the paternal and maternal mitochondria to daughter cells. Hence there is the potential for intracellular competition because there will be selection in favour of (paternal or maternal) genetic variations affecting the chance that mitochondria are transmitted to daughter cells. In particular, the competition between genetic variations affecting the chance that mitochondria have access to the cytoplasm of the gametes is important, for only these mitochondria are passed on to the next generation. This competition between mitochondrial variations may harm the host cell and the whole organism. Hence inclusive fitness theory predicts coercion suppressing the evolution of selfish mitochondrial variants. One mechanism suppressing selfish variants is uniparental (maternal) transmission of mitochondria. Assume that mitochondria are transmitted to the next generation through only the egg cell. If a selfish mitochondrial variation occurs in the male, then this variation will not be passed on to the next generation and, hence, does not affect mitochondrial evolution. And if a variation arises in the female, the variation can be passed on to the next generation, but there is no conflict here between the mitochondria and the whole organism. The probability that the genes of mitochondria and host cells are transmitted to the next generation is then similar. Hence in uniparental inheritance of mitochondria, there is no longer the potential for conflict between genes in mitochondria and genes in the cells of the organism.

Uniparental inheritance of mitochondria solves the conflict between mitochondria and the host, but it is easy to see that it creates a new conflict. For in the case of maternal transmission of mitochondria, males are for mitochondria a dead end. Natural selection will then favour genetic mitochondrial variations that shift the sex ratio towards more females, for example, by increasing the relative number of females (at the expense of the number of males). There are examples known where mitochondrial genes cause male sterility in plants. Inclusive fitness theory predicts that nuclear genes (of the host cell) will be selected that counteract the effects of these mitochondrial genes.

Conflict between mitochondria and host cells is an example of a conflict between unrelated units in an egalitarian unit. Another source of conflict is sexual reproduction. Conflict arises here because genes affecting a trait follow different transmission rules. Examples are paternally and maternally inherited genes, since the probability that they are present in offspring may be different, and genes on the sex chromosomes follow different transmission rules compared to genes on the autosomal chromosomes (see also Chapter 1, section 1.3). I will first briefly discuss the example of *sexual antagonistic evolution* and, in the remaining part of this chapter, intragenomic conflicts between paternally and maternally inherited genes.

Rice (1984) has shown that as the result of the different transmission rules of the sex chromosomes, sexually antagonistic expression of genes may evolve. Suppose that there is a recessive gene on the X chromosome, and assume that its effects are a fitness advantage for males but a fitness disadvantage for females. For example, the gene causes carriers to have small hips increasing the running capacity of their carriers. This may be an advantage for (competitive) males but a disadvantage for females because small hips increase the chance of complications during the delivery. Hence there is the potential for conflict. Because the gene is located on the X chromosome and is recessive, it can increase in frequency since the X-linked gene is at first only expressed in males (males have one X chromosome, females two). However, if the gene becomes more prevalent, the probability arises that some females in the population are homozygote and, hence, experience the disadvantage. If the cost the gene 'imposes' on females exceeds the benefits for males, Rice's model predicts the evolution of a modifier gene that reduces the expression of the allele in only females. In the end a gene evolves that is expressed only in males. Hence the outcome of this conflict is the sex specific expression of genes on the X chromosome.

3.6   Genomic imprinting

An important discovery made in the second half of the previous century is the discovery of the epigenetic control of gene expression. It was already known that the expression of a gene is regulated by factors from the environment. For example, Jacob and Monod demonstrated that a gene can be switched from an off- to an on-state by adding a chemical agent to the environment. They showed that the transcription (and the subsequent translation of mRNA into a protein) may be induced because the agent, by binding to the promoter-site of the protein synthesis machinery, alters the gene from on off-state to an on-state. The discovery that genes can be switched on and off helped us to understand how gene expression was regulated during ontogenesis (and other stages of the life cycle of a member of a species). For it was now easy to see that a similar process is in use for regulating the timing of the action of genes. If, for example, at a certain developmental stage the product of a certain gene was needed, it was conceivable that, because of the previous steps in the developmental programme, a produced protein could trigger the transcription of this gene and its production could affect the next step in the developmental programme. Epigenetics is the field in which the regulation and modulation of gene expression is studied.

One of the proximate mechanisms involved in the regulation of gene expression is DNA methylation (Kelsey and Feil, 2013; Reik and Walter, 2001). It means that especially the cytosine bases in the DNA are methylated, i.e. methyl groups are attached to bases in the DNA with the effect that expression of the gene is repressed or activated (switched off or on). The attachment of methyl groups is a reversible process, for attached molecules may be removed during later stages in the life cycle. Methylation of DNA is involved in genomic imprinting.

In genomic imprinting it is the sex of the parent that determines gene expression (see Hall, 1997; Ohlsson, 1999; Ohlsson, Hall and Ritzen, 1995; Reik and Surani, 1997; Wilkins, 2009). It brings about that the expression of a gene depends on its methylation pattern (the pattern of methyl groups attached to a gene). This methylation pattern is called an imprint. The imprint determines whether a gene is switched on or off. But in contrast to other forms of gene regulation, the environmental factor is the sex of the parent: only the sex of the parent determines whether a gene gets an imprint or not. These imprints are established during the formation of the egg and sperm cells. If, for example, a certain gene present in the gonads of a male is imprinted with the effect that the gene is switched off (and, hence, is not expressed in the offspring during a certain developmental period), then this same gene is not imprinted in the gonads of a female (and, hence, is expressed during this period). The reversed expression pattern occurs too. Imprinted genes are therefore genes that possess, as it were, information about their parental origin. The imprints are removed in the gonads of the children, and afterwards new imprints are established, depending on the sex of the child. There are 100 to 200 imprinted genes. Some researchers (Gregg *et al.*, 2010a; 2010b) have suggested that there may be more than 1,000 imprinted genes, but their findings were not verified by others (DeVeale *et al.*, 2012). The antagonistic expression of paternally and maternally derived genes in children raises the evolutionary question of why this pattern evolved.

The so-called kinship theory of imprinted genes (Haig, 2002; 2004; Mills and Moore, 2004; Wilkins and Haig, 2003) explains the evolution of imprinted genes. It is an extension of Hamilton's inclusive fitness theory. Trivers (1974; 1985) argued that Hamilton's theory can be used for explaining parent–offspring conflict. The important insight of Trivers is that investment in a certain child always has *opportunity costs*, i.e. a parental investment in the child reduces the possibility of investing in another child. Hence the more a parent invests in a particular child, the higher the opportunity costs are. Assume a very simple model, namely a family consisting of a mother and either two sibs or two half-sibs. The mother is selected to invest in a child as long as the fitness benefit (B) of the investment in a child does not exceed the opportunity costs (C) of being unable to invest the same resources in the other offspring (hence as long as B>C). The point to notice here is that the genetic relation between a mother and all her children is equal, but the relation of the offspring to other offspring is r. The value of r is ½ in the case of full-sibs and ¼ in the case of half-sibs. Hence a full-sib favours investments if B>½C, and a half-sib if B>¼C. For understanding genomic imprinting, it is important to note that in the case of half-sibs the value of ¼ is an average of a half (via the shared mother; $r_m=½$) and zero (via the unshared father; $r_p=0$). Because $r_m>r_p$, there is the potential for conflict between paternally and maternally derived genes since an act may enhance the inclusive fitness of maternally derived genes but reduce the inclusive fitness of paternally derived genes (and vice versa). Kinship theory predicts then different effects of paternally and maternally derived genes in offspring if these genes are capable of affecting maternal investments. Many studies have shown that from conception until weaning, paternally expressed genes (from now on *patrigenes*) increase the demands on the mother, whereas maternally expressed genes (from now on *matrigenes*) decrease the demands on the mother (the terms patrigene and matrigene were proposed by Queller, 2003).

What is the outcome of this intragenomic conflict? A simple model demonstrates why the intragenomic conflict results in genomic imprinting (Mochizuki, Takeda and Iwasa, 1996). Suppose that there is a conflict between paternally and maternally derived genes about the optimal level of maternal investments, and assume that this conflict concerns the size of the child. Assuming that a bigger size (as the result of more maternal investments) is advantageous for the paternally derived genes but disadvantageous for the maternally derived genes, there is an intragenomic conflict about the optimal level of a growth hormone determining the size of a child. Suppose that 5 units of the growth factor are optimal for the paternally derived genes, whereas 4 units are the optimum for the maternally derived genes. If we assume additive interactions between genes, then the model states that the paternally derived allele produces 2.5 and the maternally derived allele 2 units of the growth factor. The amount of the growth factor produced in the child is then 4.5, which is 0.5 too little for the paternally derived genes and 0.5 too much for the maternally derived genes. Natural selection will then favour an increase in the production of growth factor by the paternally and a reduction by the maternally derived genes. Table 3.2 shows that this will go on till in the end the paternal optimum is achieved. The result is genomic imprinting: the paternally derived allele is active while the maternal allele is inactive in the child. The

Table 3.2: If natural selection favours different levels of the production of a growth hormone by paternally and maternally derived alleles, then it will result in genomic imprinting: the allele favouring the highest production wins the arms race. This is called by Haig the 'loudest voice prevails' principle (Haig, 1996; 2002).

Paternal allele:	Maternal allele:	Sum:
2.5	2.0	4.5 $\rightarrow$
3.0	1.5	4.5 $\rightarrow$
3.5	1.0	4.5 $\rightarrow$
4.0	0.5	4.5 $\rightarrow$
4.5	0	4.5 $\rightarrow$
5.0	0	5.0

only possibility remaining for the maternally derived genes is to enhance the production of a growth inhibitor.

The kinship theory is also important for understanding diseases as the result of disrupted expression of imprinted genes. In Table 3.2 it is assumed that there is an adaptive interval: if the value of the produced growth hormone lies between the extremes 4 and 5, the resulting growth of the child is assumed to be adaptive. This explains the occurrence of disease symptoms in the case of diseases caused by deletions (resulting in the lack of expression of an imprinted gene) or disomies (i.e. two sets of chromosomes from a single parent). In the case of a paternal deletion, the value of the produced hormone is 0, while a disomy results in 10 units of the hormone. Both values lie outside the adaptive interval. An example is the Wilms tumour, occurring in 5 to 10% of children with Beckwith–Wiedemann syndrome (see Table 3.3). Beckwith–Wiedemann syndrome (BWS) is, just like the Angelman syndrome (AS), associated with excess expression of paternal alleles or deficient expression of maternal alleles (whereas the Silver–Russell syndrome (SRS) and Prader–Willi syndrome (PWS) are associated with excess expression of maternal alleles or deficient expression of paternal alleles; see Buiting, 2010; Eggermann, Eggermann and Schönherr, 2008). The Wilms tumour is in some cases of BWS caused by the 'double dose' of a patrigene and is therefore explicable as the result of overexpression of a paternally expressed growth hormone. Cell proliferation as the result of stimulation by a growth hormone produced by a paternally expressed gene is a normal phenomenon as long as the production lies

Table 3.3: Human imprinted disorders discussed in this book.

Human syndrome	Some causes
Angelman syndrome	Deletion of maternal 15q11-q13
Beckwith–Wiedemann syndrome	Mutation of an imprinted gene in the 11p15.5 region
Prader–Willi syndrome	Deletion of paternal 15q11-q13
Silver–Russell syndrome	Maternal disomy of chromosome 7

within the adaptive interval, but a paternal disomy results in a tumour since the stimulation of cells is then brought about by a hormone level outside the adaptive interval. Consequently, the Wilms tumour is not an 'intended' effect of a paternally expressed gene, but a side effect of a gene affecting growth processes. This is an important insight: it shows that the symptoms occurring in the case of diseases caused by deletions and disomies do not directly give us an impression of the evolutionary causes of why imprinted genes evolved. From the observed cancer we have to infer that there may be a conflict concerning growth. One can expect that in the case of the development of behaviour and the mind, it will be more difficult to infer the underlying causes of the intragenomic conflict from symptoms, especially if imprinted genes affect complex skills.

Kinship theory predicts that from conception until weaning, patrigenes increase while matrigenes decrease the demands on the mother. Úbeda (2008) has argued that the expression of imprinted genes will in humans be reversed starting with the juvenile period. Why is there a reversal in the expression of imprinted genes? The easiest way to explain this reversal is to consider the effects of dispersal. There is evidence that in the early hominids (including the Neanderthals and humans living in hunter-gatherer societies) females dispersed at puberty and joined another group. A similar pattern is observed in the other apes. If this dispersal pattern has been true for a long period of our evolutionary past, then the males in human groups were more related to each other than they are to the adult females. Females are less related to each other than males are. To put matters in a simple manner: the males in the group are mostly fathers, sons, brothers, uncles, nephews and cousins, whereas the females are immigrants from other groups. Úbeda (2008; Úbeda and Gardner, 2010; 2011) argues that this difference in average relatedness creates the potential for a conflict between paternally and maternally derived genes in offspring. This conflict concerns social behaviours affecting cooperation in the group. Paternally derived genes will favour cooperative behaviours (directed towards other group members), whereas maternally derived genes will benefit from selfish behaviours. Hence this is opposite to the pattern observed during the prenatal and neonatal stages. Note, however, that this reversal in the expression of imprinted genes only applies in the case of female dispersal. For example, in a harem structure females are more related to each other (for example, in lions, where males take over prides of lionesses). The model predicts then no reversal in the expression of imprinted genes. This prediction can easily be tested in future studies.

3.7 Prenatal and neonatal development

The kinship theory of imprinted genes predicts the pattern of genomic imprinting during the different stages of offspring development. We can use the results of empirical studies to test whether these predictions hold. From conception until weaning, paternally expressed genes increase the demands on the mother, whereas maternally expressed genes are expected to decrease the demands on the mother. In terms of inclusive fitness theory: if $r_m > r_p$, then patrigenes favour higher levels of selfishness (or lower levels of altruism), whereas matrigenes promote higher levels of altruism (and lower levels of selfishness). I discuss some examples below.

Plagge *et al.* (2004) have shown that the patrigene Gnasxl promotes the sucking response in mice via centres in the brainstem that activate muscles in the tongue and jaw. GNASxl probably has a similar function in the human species: two children who, due to a deletion, lacked the effects of GNASxl hardly sucked (Genevieve *et al.*, 2005). Another patrigene affecting sucking behaviour is MAGEL2: mice lacking the expression of the gene do not suck. Schaller *et al.* (2010) have shown that the normal phenotype can be rescued by exposure to one shot of oxytocin (five hours after parturition) in the hypothalamus. Hence MAGEL2 probably affects the production of oxytocin that functions as a trigger for the instinctive sucking behaviour.

Patrigenes that influence suckling behaviour join forces with patrigenes that modulate the use of energy gained through sucking. For example, the paternally expressed snoRNA SNORD116 in humans (also called HBII-85) and Snord116 in mice (also called MBII-85) affect growth. Mice lacking this snoRNA are of normal size at birth but fail to grow in the first three weeks when they are dependent on maternal milk. Ding *et al.* (2008) have shown that decreased milk intake is not the cause of this growth failure. They suggest that a disturbance in the growth hormone pathway is probably the cause of the reduced growth (another patrigene with a similar effect is discussed by Itier *et al.* 1998).

Kinship theory also predicts the evolution of maternally expressed genes that counteract these effects of patrigenes and contribute to mechanisms and behaviours that enhance inclusive fitness. A well-known example in mice demonstrating the counteracting effects of matrigenes is Igf2r. This imprinted gene evolved in response to the evolution of the patrigene Igf2, a growth promoter. Igf2r reduces growth since its product influences the effect of Igf2: the receptor, produced by Igf2r, is a decoy receptor that binds Igf2 and transports it to the lysosomes for degradation (Haig and Graham, 1991). An example of a matrigene enhancing inclusive fitness is the allele coding for Gαs. This imprinted gene is involved in the production of heat by brown adipocytes. In species such as mice that huddle during the preweaning stage, heat production by one individual reduces the heating costs of other individuals. An individual's heat production is therefore beneficial for the other individuals in a litter. The heat produced may be seen as a collective good that increases the inclusive fitness of the mother and the siblings (Haig, 2008; see also Haig and Wilkins, 2000; Olson, 1965). As predicted by kinship theory, the paternally derived copy coding for Gαs is not expressed and, hence, does not contribute to the common good. It acts as a 'free-rider'. This shows that the production of heat by brown adipocytes has been subject to antagonistic co-evolution of patrigenes and matrigenes. It also shows that patrigenes and matrigenes favour different allocation of energy: while patrigenes favour the allocation of energy to growth processes (enhancing the fitness of the individual), matrigenes favour the allocation to the production of heat (enhancing the inclusive fitness) during the same period.

Kinship theory also predicts that a longer duration of breastfeeding is, in the case of the human species, beneficial for paternally derived genes, since milk is probably nutritionally and immunologically superior to solid food (see Kennedy, 2005, for qualifications of this statement) and because the sucking causes lactational suppression of ovulation (and, hence, delays the arrival of a potential competitor). At face value, these objectives of paternally derived genes can be achieved quite simply, since prolongation of sucking behaviour into infancy is sufficient to accomplish these goals. By causing the release of oxytocin, sucking stimulates the production of milk and suppresses ovulation through gonadotropin-releasing hormone and prolactin. Because lactational anoestrus is a function of the frequency of nursing rather than the amount of milk produced, kinship theory predicts that patrigenes will install a pattern of regular suckling in babies by waking them up periodically (Blurton Jones and Da Costa, 1987; Haig and Wharton, 2003). There is evidence in favour of this prediction: in mice, Gnasxl is also expressed in structures in the brain that are involved in arousal (Plagge *et al.*, 2004). Another patrigene that may be involved is Magel2 (Kozlov *et al.*, 2007), for its product modulates in mice circadian rhythmicity. Interestingly, children with PWS who lack the effects of patrigenes (including MAGEL2) are somnolent and rarely awake for feeds (for reviews of PWS see Bittel and Butler 2005, Goldstone 2004), while reduced sleep and frequent waking is a feature of children with Angelman syndrome (AS; lacking the expression of matrigenes: for reviews of AS see Clayton and Laan, 2003; Williams *et al.*, 2006). Another relevant observation is that breast milk contains sedative substances such as benzodiazepines (Dencker *et al.*, 1992): it is possible that these substances oppose effects of patrigenes (see further Haig, 1993).

3.8 Sibling rivalry and cooperation

Imprinted genes also affect brain and behavioural development during neonatal, juvenile and later stages. Kinship theory predicts that patrigenes favour selfishness and matrigenes altruism as long as children are dependent on maternal investments, and predicts in humans a reversal in this pattern after the child's sustenance shifts from family to group. Investigating the possible effects on children's brain and behaviour development is also interesting because of the development of the mind. According to the neo-Aristotelian conception the mind develops out of instinct as the result of learning and teaching. This raises the problem of whether and how imprinted genes affect the transition from instinctive to typical human, rational behaviour. I shall discuss possible effects of imprinted genes on the development of linguistic behaviour in Chapter 7 and on moral behaviour in Chapter 8. I discuss the transition from breastfeeding to consuming solid food in this chapter.

The instinctive sucking movements are generated through the brainstem and are, at first, not 'controlled' by the hypothalamus and the cortex. This is inferred from the fact that in mice, the sense of taste (indicated by decreased acceptability of quinine over water), hunger and satiety mechanisms are largely absent before weaning (Henning, 1981). The transition towards adult mechanisms occurs in mice during weaning (after about three weeks). Weaning in mice is a short period and marks the beginning of a life independent of nursing. Permanent molars develop and eyes and ears open just before pups begin the exploration of their environment. Hence in mice, the transition from sucking behaviour to foraging for solid food is the transition from instinctive suckling to volitional eating and foraging behaviour.

In humans, infants younger than three weeks of age do not respond to salty or bitter adulterations of their formulas (although eight-day-old infants are sensitive to the smell of their mother's breast pad, see Blass and Teicher, 1980). Just as in mice, the internal mechanisms involved in hunger and satiety are absent during these early stages of human development. In infants younger than twelve weeks of age sucking terminates with sleep and is reinstated with awakening. In children older than twelve weeks, sucking is not terminated by sleep: children start to play with the nipple or bottle after a feed. The child's attention and motor patterns are no longer dominated by a pattern that culminates in sucking. Children start to interact with their parents and the sucking behaviour gradually develops into volitional behaviour. When infants learn to express their needs and wants verbally during the second year, this volitional behaviour is extended with linguistic behaviour. The child develops then primitive forms of intentional behaviour, for he or she can then ask for the breast if he or she is hungry.

The evolution of the transition from instinctive to volitional and intentional behaviour in humans is linked to a developmental pattern that evolved only in humans: humans started to use protein-rich meat (but also plant food and tubers) as a supplement to and alternative for maternal milk. Lactational anoestrus was therefore removed at an earlier age. The result was that humans had a shorter interbirth interval than our closest relatives, but a longer juvenile dependence (Bogin, 1997; Bogin and Smith, 1996; Kennedy, 2005; Humphrey, 2010). Humans wean in natural fertility populations at about 1 to 4 years, while chimpanzees wean at 5 and orang-utans at 7.7 years. Hence in contrast to chimpanzees, weaning in humans does not mark the transition to an independent life, for they remain dependent on their mother for supplementary food. Because mothers had then to raise several children in the same period, competition between human siblings was intensified. Data supporting this hypothesis are, for example, studies in Malawi and India showing that child mortality increases as the interbirth interval decreases (Manda, 1999; Shahidullah, 1994). Since infant mortality has been high during the largest part of human evolution (as it nowadays still is in apes: 40 to 50% die in infancy; in human foragers it is estimated that 30 to 40% of the children die, see Sellen 2007), sibling rivalry was a substantial evolutionary force. This observation raises the question of what the opportunities were for patrigenes and matrigenes to affect parent–child and child–child interactions during postnatal stages. Assuming that competition between siblings is beneficial for patrigenes but disadvantageous for matrigenes, and that offspring have been dependent on breastfeeding, supplementary food and care, kinship theory predicts that matrigenes promote a fair sharing of food among siblings, while patrigenes promote behaviours that will help an individual to compete with siblings. I discuss here one example: the possible effects of patrigenes on crying and reactive crying (and discuss other possible effects of patrigenes and matrigenes in Chapters 7 and 8).

It is assumed that crying enhanced maternal care and, hence, is probably affected by patrigenes (see Brown and Considine, 2004; Trivers, 1974; see also Horsler and Oliver, 2006). Interestingly, newborn babies also cry when they hear another infant cry. Simner (1971) found it in two- and three-day-old babies and noted that it was not the loudness of the cry that evoked the response. These findings have been replicated by Sagi and Hoffman (1976; see also Hoffman, 2000) in one-day-olds. They showed that it is not a simple imitative vocal response lacking an affective component. The reactive cry is indistinguishable from the spontaneous cry of an infant who is in actual discomfort. Martin and Clark (1982) showed that infants do not react as much to the sound of their own cry. This finding has been replicated by Dondi, Simion and Caltran (1999). Psychologists (e.g. Warneken and Tomasello, 2009) often state that reactive crying is an example of empathic behaviour showing that humans are natural altruists (see the discussion in Smit, 2010d), but Trivers' theory predicts that this innate response to the cry of another of the same species is an 'egoistic' response since it increases the chance that the child receives maternal care. Hence we expect patrigenes to favour this response.

After six months reactive crying appears to decline: 6-month-old children responded to distress of another only after the other displayed several instances of distress. The cry displayed by a 6-month-old is different from a newborn's cry: the infant looks sad and puckers his lip before starting to cry, just as infants do when they are in actual distress. This fact was already noted by Darwin (1877, p. 289) who described the empathic response of his 6-month-old son to his nurse pretending to cry by 'his melancholy face, with the corners of his mouth well depressed'. Hence 6-month-old infants no longer respond 'mechanically' to another's cry. One-year-old children still respond to the distress of another by displaying distress behaviour themselves. Yet the response of a 1-year-old differs from the response of a 6-month-old. Children respond by looking sad, pucker their lips and then start crying, but their cry is now accompanied by whimpering and silent watching or staring. Moreover, as soon as children are able to crawl and later become truly self-moving creatures, they actively seek comfort in their mother's lap. At about 14 months old children begin making helpful advances towards the distressed other through patting and touching, which soon gives way to more differentiated positive interventions such as kissing, hugging, giving physical assistance, getting someone else to help and giving advice and reassurance. Yet although children sometimes help others during the second and third year, the base rates are low (Svetlova, Nichols and Brownell, 2010). Furthermore, helping is then circumstance-relative: 2- and 3-year-olds sometimes help when they witness a distressed other, but do not help if they cause the distress (see Zahn-Waxler, Radke-Yarrow and Wagner, 1992).

3.9   Grooming and caching food

Investigations of effects of imprinted genes show that these genes are involved in *trade-offs* in fitness. I have already discussed some examples: the trade-off between a muscular body and storing fat; between growing tall or staying small; between helping and being selfish. Kinship theory predicts that patrigenes and matrigenes weigh the benefits and costs of these trade-offs differently and have therefore different effects. Some of these trade-offs concern social behaviour: between egoistic behaviour and helping another; between impulsiveness and executive control. Trade-offs concerning the brain are interesting for they may help us to understand the evolution of animal and human behaviour better. I shall discuss two examples.

Haig and Úbeda (2011) have argued that self-grooming and allogrooming have been subject to genomic imprinting. Self-grooming and allogrooming (a form of helping behaviour: removing lice) are part of social hygiene for they reduce the spread of ectoparasites in the population. Hence self-grooming and allogrooming benefit the survival of individuals and the group and the evolution of these behaviours is explicable in terms of inclusive fitness theory. As always, there is the potential of a free groomer, for an individual that does not groom others but is being groomed by them does not pay the costs of allogrooming. A free groomer has therefore the possibility to use the saved energy for other activities (e.g. barbering; see below). Because of this trade-off between grooming and other activities, kinship theory predicts that patrigenes and matrigenes have different interests.

There is an interesting link between grooming and the immune system which was discovered by accident in 2002. When studying the Hoxb8 mutant, it was noted that these mice exhibited compulsive grooming and fur loss with 100% penetrance. The mutant mice engaged in excessive self-grooming as well as grooming of wildtype cage mates. This was unexpected given that Hoxb8 is a member of a large family of transcription factors best known for their roles in establishing body patterning during embryonic development. This observation raised the question why a mutation of Hoxb8 causes obsessive–compulsive grooming behaviour.

It was found that the link between Hoxb8 and grooming was microglia cells. These cells originate in bone marrow and migrate during the post-natal stages to specific brain areas of the striatum and then regulate grooming behaviour. Grooming in mice is a fixed action pattern with what is called a syntax (first the head, then the body and then the tail). The Hoxb8 mutant showed that dysregulation of grooming by microglia cells leads to obsessive–compulsive grooming. Interestingly, the obsessive–compulsive grooming behaviour of mice can be rescued with a bone marrow transplantation (of normal microglia cells), because bone marrow is the source of these cells. Since microglia cells are (evolutionarily) derived from the macrophages (involved in the immune response), grooming regulated by microglia cells showed an interesting link between social hygiene and the immune system (see also Ting and Feng, 2011).

There is evidence that imprinted genes may be involved here, for children with PWS (lacking the expression of patrigenes) display obsessive–compulsive skin picking. It is known that there is a locus linked to grooming adjacent to the imprinted gene cluster associated with PWS. Hence it is possible that their obsessive–compulsive behaviour is caused by the lack of the expression of a patrigene in the striatum.

The possibility that there is a trade-off involved here between grooming and barbering is based on studies of the effects of the imprinted gene Grb10. Grb10 is expressed from the maternal allele in foetal tissues (and is a growth inhibitor), but expressed from the paternal allele in the brain. Mice with an inactivated paternal copy in brain tissues are more aggressive (they rarely backed down in a test in which they and a control entered a narrow tube from opposite ends) showing that GRB10 indeed affects brain and behaviour during later stages. These mice barbered the whiskers (they pluck them) of cage mates more often than controls. There is a link here with grooming, because barbering occurs during bouts of mutual grooming. Barbering is probably a signal of dominance, and there is evidence that the barbered mouse backs down when confronted in the tube test by their dominant barbers. Although barbering is displayed during grooming, these behaviours appear to be regulated by different genes since inactivation of genes involved in grooming genes does not affect barbering. The Grb10 gene is involved in SRS (the opposite of BWS).

Haig and Úbeda (2011) discuss two possible explanations of differential effects of paternally and maternally expressed genes here. First, it is possible that greater investment in self- and allogrooming (as hygiene-related behaviours enhancing the survival of the group) benefit maternal alleles if the individuals are more related through their maternal genome. This would explain why patrigenes reduce the investment in allogrooming and increase social dominance. Second, social dominance is associated with a less fair sharing of resources and this would also explain why the paternally expressed genes favour an increase in frequency of dominance-related behaviour. However, if the primary effect of Grb10 is to reduce the assertion of dominance-related behaviour through barbering, then domin-ance relations are predicted to have involved individuals who were closer relatives on the paternal side.

Another example where imprinted genes may affect trade-offs involved in brain and behaviour development is inferred from the observation that children with PWS selfishly hoard food. The results of studies in birds show why there are trade-offs involved between hoarding or caching food versus 'stealing' food as a free-rider. Caching food is comparable to storing fat: in both cases food is stored up in times of plenty with the aim of using it later in times of scarcity. Bears consume more food during autumn and store food as body fat, used when they hibernate during the winter, whereas birds like titmice cache seeds in scattered places in the autumn and retrieve them during the winter and spring. In some titmice an individual bird may store 100,000 to 500,000 seeds during the autumn in separate places. One can easily imagine the problem of a free-rider here. Suppose that an individual does not cache or hoard food and, hence, does not pay the costs of caching. Then this individual could benefit from the stored seeds of others. Inclusive fitness theory predicts mechanisms that reduce the costs of free-riding. Brodin and Ekman (1994) have shown that storing birds benefit more from their hoarding behaviour: they were about five times more likely to recover their items than other birds in the same group. This shows that memory is responsible for the benefit of caching. Studies also show that birds that store food have a larger relative volume of the hippocampus, a brain structure involved in memory. There is also evidence that the volume of the hippocampus is reduced in the summer, probably because maintaining a larger volume is costly. Hence caching food and recovering involves an investment cost in memory which the free-rider does not have. This possibility raises the question whether abnormal hippocampal functioning seen in children with PWS is linked to this trade-off (see further in Chapter 8).

3.10 Conclusion

I have elaborated in this chapter inclusive fitness theory and one of its applications: the kinship theory of genomic imprinting. Hamilton's theory, we can conclude, is a very powerful theory capable of explaining many phenomena at several levels of organization. A similar conclusion applies to the kinship theory of genomic imprinting: this theory clarifies why the expression of some genes depends on the sex of the parent and why imprinted genes have key roles in resource transfer from parent to child.

In Chapters 7 and 8 I will apply inclusive fitness theory to typical human features, namely language evolution and the evolution of moral behaviour. I shall also discuss in these chapters possible effects of imprinted genes on these features. Since we then enter the domain of psychology and, hence, are confronted with the problem of how the human mind evolved, I will use the conceptual insights discussed in Chapter 2. Evolutionary theorists expect that inclusive fitness can also be applied to the evolution of the human mind and have argued that there are trade-offs concerning mental functioning (see Badcock and Crespi, 2006; 2008; Haig, 2006; 2010; Úbeda and Gardner, 2010; 2011; Wilkinson, Davies and Isles, 2007). Examples they mention are possible trade-offs between immediate and delayed gratification, between empathizing and systemizing, between focused and diffuse attention and between impulsiveness and executive control. A key challenge for establishing a synthesis between inclusive fitness theory and psychology is, according to them, to understand how psychological trade-offs mediate life-history trade-offs. Given what I have argued in Chapter 2, it is no surprise that I will argue that the success of this synthesis depends on whether we use the Cartesian or Aristotelian conception of the mind. The challenge we face is to connect inclusive fitness theory to the neo-Aristotelian conception in such a manner that the resulting conceptual framework is capable of explaining language evolution, moral behaviour and features of the human mind.

In Chapters 4, 5 and 6, I shall elaborate the Aristotelian framework extended with inclusive fitness theory. Chapter 4 discusses the relationship between teleology and evolutionary theory. This is an important issue, because many evolutionary theorists assume that Darwin's critique of the argument from design implies that natural selection replaced teleology.

I shall argue that this idea is mistaken: Darwin eliminated the concept of design, rehabilitated Aristotelian teleological explanations and extended the Aristotelian framework with evolutionary explanations. He demonstrated that we can explain ends and purposes in nature without referring to the wisdom and power of a Designer. The advantage of the 'correct' Darwin interpretation is that it helps us to reconcile evolutionary theory with medicine, psychology and law.

I have explained in Chapter 2 some differences between the Aristotelian and Cartesian conception of the mind. Darwin's rehabilitation of Aristotelian teleology enables us also to develop a modern version of the Aristotelian concept of the mind. In Chapter 5 I shall discuss the neo-Aristotelian conception and will confront it again with the crypto-Cartesian one. Modern materialists try to explain the evolution of the human mind as the product of brain states or processes. Neo-Aristotelians, by contrast, argue that the human mind evolved out of animal behaviour as the result of language evolution. I shall elaborate the differences between these two conceptions through discussing the ideas of crypto-Cartesians and neo-Aristotelians about behavioural flexibility and conflicts in our minds.

In Chapter 6 I use the neo-Aristotelian framework extended with Darwin's theory for solving and resolving some problems concerning the evolution of human cognition. It is believed that there is an essential difference between animal and human cognition because only humans are culture-creating creatures. This has freed our species from its primate heritage, for natural selection was extended with cultural selection. This observation raises the problem of how human cognition evolved. Some evolutionary theorists have argued that for understanding human cognition, we have to return to Lamarck's theory, because adaptive evolution started, according to them, with adaptive learning. Through extending Weismann's critique of Lamarck's theory with some conceptual insights (first discussed by Aristotle and later elaborated by Wittgenstein and others), I shall argue that these ideas of neo-Lamarckians are highly problematic.

CHAPTER 4

Evolution, teleology and the argument from design

4.1 Introduction

Descartes once argued that we should consider God as the efficient cause of all things. Studying the mechanical world will provide us with material to glorify the great Designer. He advised against the use of teleological explanations, since studying goals in the living world shows that humans are arrogant to suppose that they can share in God's plans (see further Hacker, 2007, chapter 7).

Darwin was familiar with the critique of Cartesians on the use of (originally Aristotelian) teleological explanations (Lennox, 1992). He real-ized that the development of the theory of evolution enables us to return to teleological explanations in animate nature. According to Darwin biolo-gists should no longer explain goals and purposes in nature as extrinsic parts installed by a Designer (as was argued by Paley, 2006 [1802]), but should study them as intrinsic parts (as was argued by Aristotle). Darwin's rehabilitation of teleological explanations was already noted by Asa Gray (1874, p. 81) in his tribute to Darwin. He remarked that we should 'recognise Darwin's great service to Natural Science in bringing back to it Teleology: so that instead of Morphology *versus* Teleology, we shall have Morphology wedded to Teleology'. Darwin's response to this remark is telling: 'What you say about Teleology pleases me especially, and I do not think any one else has ever noticed the point. I have always said you were the man to hit the nail on the head' (Darwin, 1887, p. 189). Evolutionary theory can help us to explain the origin, maintenance and modification of goal-directed phenomena. For instance, evolutionary theory explains why during the evolution of large animals a heart evolved, since diffusion was no longer adequate to distribute nutrients and oxygen through the body. The heart evolved to serve a function, i.e. it exists for a goal. Teleological explanations clarify why a normal functioning heart contributes to the health of the organism and why a malfunctioning heart results in disease.

Darwin's rehabilitation of teleological explanations and elimination of the concept of design are not the standard interpretation of Darwin's ideas among evolutionary biologists. On the contrary, Ruse (2003, p. 266; see also Ayala, 2007) has observed that 'the metaphor of design, with the organism as artefact, is at the heart of Darwinian evolutionary biology'. An important source of this idea is Williams' *Adaptation and natural selection* (1996 [1966]). Williams thought that Darwin replaced teleology by natural selection and argued that natural selection has design-creating capacity. In this chapter, I shall discuss the differences between these two views and argue why we should prefer the idea that Darwin rehabilitated teleological explanation. My argument is that it leads to more interesting and promising explanations of phenomena and helps us to reconcile evolutionary theory with medicine, psychology and law. I shall mainly discuss medical examples in this chapter, and shall discuss implications for evolutionary psychology in Chapter 5. In Chapters 7 and 8 some empirical consequences are discussed.

4.2   The return of teleology

Aristotle distinguished four causes: material, formal, efficient and final. These causes explain why things are as they are (see Aristotle, *Physics* 195a3, in Barnes, 1984, pp. 315–446). For example, a hammer is made of metal and that is its material cause. The formal cause is that the hammer has a handle and a head. The efficient cause is the construction of a hammer by a craftsman. The final cause is what a hammer is made for: it is made for hammering. The four causes can also be distinguished in the case of an organ of a living being. The material cause of an organ like the heart is the muscle cells; the formal cause is the subdivision of the heart into ventricles and atria and how these are connected. The heart evolved during the development of an individual by cell division and differentiation. It has also undergone an evolutionary history. Both the ontogenesis and the phylogenesis belong to the efficient cause. When the heart has matured in the embryo (and later the foetus), its contractions are rhythmically generated by nerves. This explanation of the heart's activity is also part of the efficient cause. The heart pumps blood through the vessels and this is the final cause. This function of the heart was discovered by William Harvey. This discovery shows that we do not always know the final causes of organs in advance of investigation. It also demonstrates that answers to the question, 'What is the function of an organ?', are independent of data about the course of the ontogenesis and phylogenesis of organs.

Descartes criticized the Aristotelian framework and stated that there are no goals in nature. He objected to the use of final causes (nowadays called teleological explanations) for understanding phenomena. Descartes' critique of Aristotelian final causes is, till this day, a source of confusion. I discuss two reasons why there is confusion.

First, confusion arose because of Aristotle's ideas about the explanation of phenomena in the inanimate world. Aristotle also thought that there are goals in the inanimate world. If a gas (that consists according to Aristotle of 'fire', a warm, dry element) ascends then, according to Aristotle, it moves to its destination. And if a stone (that consists, according to Aristotle, of 'earth', a dry element) falls down then it moves to its natural location. In the seventeenth century physicists discovered that we can explain these phenomena without reference to goals or final 'becauses'. The scientific revolution led to the conclusion that the movements of bodies in the natural world are fully understandable in terms of the models of mechanics. Since some phenomena in the living world may be explained in terms of mechanical models too, the question arose as to whether teleology can be eliminated in life sciences as well. If, for example, an apple falls down from the branch of a tree, then looking for 'final becauses' makes no sense to us (we do not even understand why Aristotle appealed to teleological explanations here). However, there are also examples where it makes sense to use teleological explanations for understanding phenomena (and they are used by biologists and medical doctors). The function of organs in organisms is an obvious example. Organs, like machines, have an internal organization (the heart is composed of ventricles, atria and valves, which are linked). Since organs are akin to machines in this respect, their functioning may, in part, be explained in terms of models from mechanics. For instance, a description of the blood flow though the ventricles and atria in the heart resembles the description of a pump, since we use the same concepts in both explanations (like mass, velocity and pressure). If we describe how the blood flows from the left atrium to the left ventricle, and afterwards from the left ventricle to the aorta, then the connection of the valve to the inner walls (via the *chordae tendineae*) of the ventricle explains how the valve prevents blood flowing back to the left atrium. The valves allow blood to pass from the atrium to the ventricle, but the pressure that arises when the ventricle contracts closes the left atrium off. The *chordae tendineae* prevent the valve from turning over. Hence blood is always pumped out of the ventricle into the aorta. Since we can describe this part of the heart's functioning in terms of mechanics, the heart is akin to a machine. Nevertheless there are clear differences between an organ and a machine.

An organ has a function, i.e. its activity contributes to the health of the organism, whereas a machine is designed for a purpose defined by us. And because an organ contributes to the good of a being (whereas a machine has no good), only organisms can suffer from a disease of an organ, experience sensations such as pain and fatigue, etc. If teleology is eliminated from scientific discourse, then it is impossible to talk about goal-directed behaviour and it becomes hard to talk about health and disease. How should one explain the health and disease of organisms and symptoms of disease if one cannot refer to the function or malfunction of organs?

Second, confusion arose because of the introduction of the argument from design by Thomas Aquinas (see Lennox, 1992). While in the ancient, Aristotelian philosophy goals were seen as intrinsic parts of nature, they were explained by Aquinas as extrinsic parts of nature. Naked purposes were dressed up with the argument from design (see Kenny, 1988a). According to the medieval philosophers, there was a Creator who created living creatures according to a design. So the apparent goal-directedness in nature was seen as the product of a Designer, just as the order and harmony in the inanimate world were seen by Galileo and Descartes as the result of His work and revealed to us in the language of mathematics.

Philosophers like Descartes underpinned the argument from design with an extra argument. Descartes stated that we should not be so arrogant as to suppose that we can share in God's plans (see Descartes, 1985b, p. 202). We should consider Him as the efficient cause of all things and that studying those effects will provide us with enough material to glorify Him through examining the mechanical world. Yet do these arguments demonstrate that we cannot know the purpose of, say, the wings of a bird or the role of the heart in the circulatory system? The answer is no, for the argument of Descartes is an example of what logicians call a non sequitur. It does not follow from the premises that there are no knowable goals in the living world. All that follows is that, assuming the Cartesian premises are true, we may not be able to guess the plans and purposes the Creator had when he created animal species. However, why should we take these premises as a starting point when we are studying health and disease in the living world?

The ideas of Descartes have dominated Western philosophy for a long period. They resulted in popular discussion about the problem of whether animals are machines without a mind and humans machines with a mind. Only after Darwin formulated his theory of evolution and eliminated the argument from design, did it become possible to return to the alternative Aristotelian conception and to rehabilitate the naked conception of goals.

The concept of design was eliminated in life sciences following the development of Darwin's theory of evolution. Darwin had defended the argument from design during his youth since he was influenced by Paley. This highly influential natural theologian had argued in 1802 that the adaptiveness of forms of life demonstrate that there must be a Designer (Paley, 2006 [1802]). Darwin discovered that we do not need a Designer in order to understand the goal-directedness in the living world. In his autobiography he wrote:

> The old argument of design in nature, as given by Paley, which formerly seemed to me so conclusive, fails, now that the law of natural selection has been discovered. We can no longer argue that, for instance, the beautiful hinge of a bivalve shell must have been made by an intelligent being, like the hinge of a door by a man. There seems to be no more design in the variability of organic beings and in the action of natural selection, than in the course which the wind blows. (Darwin, 1993 [1958], p. 87)

Darwin eliminated the argument from design but did not eliminate teleology, as some scientists and philosophers in the past have thought, because of the misleading connection between designs and goals in the Judaeo-Christian tradition.

What are the differences between the Aristotelian conception of the four 'becauses', the mechanical world picture of Descartes and the picture that Darwin enabled us to develop? The difference between the post-Darwinian picture and the Aristotelian one is that we can subdivide the efficient 'because' into two or three forms of explanation (ontogeny, phylogeny and here and now). Compared to the original Aristotelian picture there is an *extension* with evolutionary explanations. The difference between the mechanical world picture and the post-Darwinian one is that we explain the goals and purposes in the living world as intrinsic parts of that world, not as extrinsic ones installed by a Designer. Research into the goal or function of phenomena provides answers to the question as to what an organ or a species-specific activity is for. In short, Darwin enabled us to study the evolutionary explanations of phenomena *and* to return to the original, naked, Aristotelian conception of purposes in just the living world.

4.3 Types of teleological explanations

I distinguish three different types of goal-directed phenomena related to different levels of organization (see Figure 4.1; for previous discussions of teleology and evolutionary biology see Mayr, 1992; Ayala, 1970). *First,*

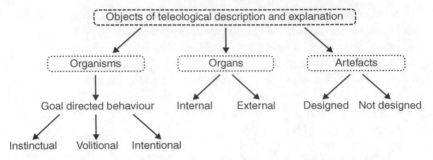

Figure 4.1: A classification of the objects of teleological explanation.

organs have a function. Malfunctioning organs have causal effects resulting in diseases and diminish, therefore, the chances of survival (and, hence, may reduce the fitness of the individual). The evolution of organs in large organisms is thought to be understandable in terms of inclusive fitness theory. A first step towards an organism with organs was that cells started to cooperate because they were kin. Cooperation can easily develop if the cooperating cells are clonally derived from a single cell, for such a bottle-neck guarantees that the cells are identical (r=1 if there are no mutations). A similar bottleneck explains why females are the cooperative part of insect societies: these societies are often started by a single queen fertilized by a single male and this guarantees that the genetic relatedness among the female insects is three-quarters. Kirschner and Gerhart (2005, pp. 35 ff.) have argued that the development of an epithelium, a closed sphere of cells, was another key element. The cells of the epithelium have junctions so tightly wedded together that virtually nothing can pass between them. Pumps and channels evolved that controlled the salt composition of the internal milieu. The controlled internal milieu facilitated further commu-nication between cells resulting in functional specialization. However, since division of labour and functional specialization are products of multilevel selection, conflict and mediation of conflict among cells have played a role during the transition from unicellular organisms to organisms with organs and tissues (Michod *et al.*, 2006; Okasha, 2006). For example, when multicellular organisms (without organs) developed a structure for storing resources, this cooperative trait created the potential for conflict, since individual cells that invested fewer resources into the structure would have more resources remaining for reproduction. Hence suppression of competition of cells was a second step during the transition. Conflict suppressors reduced the potential for within-group change, but it may also have enhanced the power of between-group selection and, hence, the development of adaptive traits enhancing the chances of survival of the whole collective. The outcome of this third step was therefore that the fitness was further 'exported' from the level of the cell to the collective.

The result of the transition from unicellular to multicellular organisms with organs was that cells and organs became parts of a being that had or would evolve its own purposes. Hence, the purpose of organs became subservient to the good of the being whose organ they are. Note that machine parts also have a purpose and these parts are also subservient to the functions of the machine of which they are parts – but not to its good, as it has no good (von Wright, 1963). A machine is created and used for a purpose defined by us. Also notice that, while we can discuss the question of whether cells or insect colonies are organisms (see Queller, 2000; Strassmann and Queller, 2010), it makes less sense to ask whether organs are organisms (is my heart an organism or only an organ?).

Malfunctioning organs have causal effects (e.g. a malfunctioning heart causing angina pectoris) and this data is used by medical doctors when they investigate diseases. If two persons have a sensation (pain) in corresponding locations in the body, and if this sensation has similar phenomenal qualities, then for medical doctors these data are an argument for investigating the possibility that these persons have the same disease. That well-functioning organs are partially constitutive of the health of organisms also explains why we do not apply the concepts of health and disease to unicellular creatures (such as bacteria) and multicellular organisms without organs (such as a colony of yeast cells), although there are, of course, disturbances in these creatures. For example, a colony of yeast cells grows abnormally, and this is caused by dysfunctioning mitochondria (the so-called 'petite' mutation). As dysfunctioning mitochondria cause diseases in humans, one may ask whether the colony is sick. The colony does not have internal and external organs and cannot express its sensations – pain, fatigue – in non-verbal and verbal behaviour. Moreover, the colony cannot report on the loss of abilities (paralysis, blindness, etc.). Hence, there are no reasons to apply the concepts of health and disease here, for all we observe are growth problems. By contrast, it makes sense to apply the concept of reproductive fitness to both an organism with organs and a colony. Furthermore, an evolutionary biologist will study the fitness effect of the 'petite' mutation at two levels. The 'petite' mutation allows faster replication of the mitochondrial genome (with the result that these mitochondria outcompete mitochondria without mutation), yet reduces the fitness of the whole colony because of the impaired growth (Taylor, Zeyl and Cooke, 2002).

The *second* type of phenomena is of self-moving organisms that, using senses, develop goal-directed activities. These are called goal-directed because there are (species-specific) regularities discernable in these activities (they consist of a concatenated sequence of behaviours ending with what ethologists have called a consummatory act). Some of these are non-cognitive (and largely innate) instinctive behaviours (Tinbergen, 1989 [1951]; Chapter 6). These behaviours are thought to have been selected because of their (individual or inclusive) fitness-enhancing effects. If these behaviours are non-cognitive (and in the case of humans not amenable to reason), then they can be understood in terms of a combination of teleonomic and causal explanations (teleonomic explanations mean that regularities in activities are explicable in terms of purposes). The sucking of a baby is an example: the child seeks about with its mouth and this seeking behaviour ends with a consummatory act. However, as noted by Weismann (1983 [1904], vol. 1, p. 143), if the sucking 'is continued into the second year of life', it is an example of volitional behaviour because the 'child knows exactly why it wants the breast'. The child can express its bodily sensations verbally and can ask for the breast if it is hungry. It no longer seeks the breast instinctively but acts in conscious response to its bodily desire, i.e. hunger or thirst. In addition, if it acts out of a desire, the child learns to refrain temporarily from satisfying the desire. As this extension of instinctive behaviour is, in part, based on learning, it is assumed that the resulting behavioural flexibility has adaptive value. Kinship theory predicts conflicts between parent and offspring about the timing of the transition from instinctive to volitional and intentional behaviour (Trivers, 1974; Smit, 2002; 2005; 2006; 2009).

Third, in the case of humans, intentional behaviour is distinguished as a distinct form of goal-directed behaviour. Intentional actions are goal-directed behaviours for which human beings, as language-using creatures, can give reasons (hence, we can develop so-called ideographical explanations for these behaviours). These actions are, in part, linguistic extensions of natural goal-directed behaviours (also displayed by other animals, such as exploring the environment) and develop when children learn to give reasons for their (volitional) plans and actions through answering what- and why-questions asked by parents (e.g. 'What do you want to achieve?'; 'Why are you doing that?'). They also then learn to account for what they are doing or intend to do, or have done.

The goal-directed behaviour is not always explained by reference to the future event or state at which an organism aims (Kenny, 1988a). The goal-directed activities of a bird building a nest are directed towards a final state:

a ready-to-use nest. However, a bird that is threatened or attacked by another will flee: it wants to move away from the location where it is, and this wanting-to-leave-the-spot may then be described as the purpose of its activity. In humans, intentional behaviour may be explicable in terms of both prospective goals and past events. Suppose someone acts out of revenge or is jealous, then he may be motivated to do something in the future because of something that has happened in the past. An example is a soccer player who has the intention to foul an opponent because he has been tackled earlier during the game.

Intentional explanations are crucial for forensic purposes to establish whether persons can be held responsible for their deeds. Since animals are not language-using creatures, they are not amenable to reason and intentional explanations are, therefore, not applicable to them. If, for example, a group of lions attacks a zebra, we hesitate to say that this is an example of a planned, intentional act. For although lions do have purposes and are able to develop primitive plans, they cannot give reasons for these plans. Hence, lions cannot be said to be accountable for their planned behaviour and we cannot call lions up to a tribunal because of their 'intentional killings' of zebras, or, in the case of males, because they have killed the young when they took over a pride (as if the male has committed a crime). If we do apply the concept of intention to animals, then we call their intentional behaviour simple or primitive, because the primary application of this concept is to a language-using being.

4.4 Williams' argument from design

A characteristic is an adaptation if it is shown that the characteristic increases the reproductive success of the organism carrying the gene correlating with the characteristic. Not every characteristic is, however, an adaptation, for the characteristic may also evolve as the result of neutral evolution.

Some one-way powers are examples of adaptations and it is useful to differentiate first-order from second-order powers. An example of a first-order power is the functional production of lactase by bacteria as a response to the presence of lactose in the environment (Jacob and Monod, 1961); an example of a second-order power is the development of an immunological memory after exposure to a certain antigen (Jerne, 1955; Burnet, 1957). These examples show that organisms are endowed with one-way powers and only respond to an environmental stimulus if it results in improved survival or reproduction. If the stimulus is absent, the adaptation imposes a cost and, hence, is not expressed. The goal-directed activities may have evolved as adaptations and this explains why mutations affecting these activities reduce reproductive success (e.g. as the result of malfunctioning organs affecting health).

Teleological and evolutionary explanations are two different yet complementary types of explanation. Two reminders are relevant here. First, teleological explanations are not causal explanations. Aristotle used the term 'aition', and it is better to translate his use of this term as the final 'because' or form of explanation (see Aristotle's *Physics*). Second, although teleological explanations are not causal efficient explanations, malfunctioning organs have causal effects. This explains why the symptoms of malfunctioning organs for medical doctors are a reason for saying that there must be a cause for the disease. By contrast, a normal, functioning organ does not require a special explanation. Hence, when we speak of a healthy, normal organ, we utter a normative expression or proposition, because a normally functioning heart is good for the organism (see von Wright, 1963, chapter 3). However, this normative expression is not a (culture-relative) value statement (see further below).

Williams argued that natural selection should replace teleology. He coined the term 'teleonomy' for the evolutionary study of functional adaptations (hence, he changed the Aristotelian meaning; see above). According to Williams 'the term [teleonomy] would connote a formal relationship to Aristotelian teleology, with the important difference that teleonomy implies the material principle of natural selection in place of the Aristotelian final cause' (1992 [1966], p. 258). This leads to the well-known conclusion that a full understanding of phenomena requires a combination of proximate-causal and ultimate-causal explanations (for the difference between these two causal explanations, see Mayr, 1961; Tinbergen, 1963). There is no space left here for teleological explanations. In order to understand Williams' view, it is useful to contrast his ideas with those of Descartes. Like Williams, Descartes believed that organisms are complex machines. However, in contrast to Williams, Descartes argued that we should not invoke the concept of a goal when explaining phenomena. For example, he argued that we could understand the working of an organ, such as the heart, without the concept of a purpose (Kenny, 1968, chapter 9). While Harvey had argued that a contraction of the ventricle squeezed the blood into the aorta, Descartes (1985a, pp. 135–139) held that blood left the heart in diastole: it was expelled into the arteries because it was rarefied by a 'dark fire' in the heart, that is, a fire similar to the one that makes hay hot when it is shut up before it is dry. Descartes preferred his theory on a priori grounds because his explanation was, in contrast to Harvey's explanation, mechanistic. His explanation required nothing but heat, rarefaction and expansion, properties also to be found in inanimate things. In Harvey's explanation the ventricle is treated as a muscular sac and the heart is said to have a function in the circulation of the blood. Descartes recommended his theory to those 'who know the force of mathematical demonstration', but biologists prefer Harvey's theory on empirical grounds.

Williams did not adhere to the Cartesian idea that explanations should be purely mechanistic. He used the argument from design for understanding why biologists 'recognize' functional adaptations. Williams argued that the demonstration of design provides the answer to the teleonomist's prime question – how he recognizes adaptations: 'functional design is something that can be intuitively comprehended by an investigator and convincingly communicated to others' (1992 [1966], p. 260). However, Williams' replacement of teleology by natural selection leads to conceptual confusions and misunderstandings. I shall discuss three of them.

4.4.1 Anthropomorphism

Williams replaced teleology by natural selection and used the argument from design for explaining functional adaptations. Yet evolutionary biologists also notice that biologists, medical doctors, psychologists and lawyers talk about goals, purposes, ends and intentions. Hence the problem they experience is if and how their models license the use of these concepts. Since there is no Designer, and because teleology is replaced by natural selection, there appears to be only one alternative left: the attribution of goals to nature is a metaphor (Ruse, 2003) or based on an analogy with human intentional actions. For example, Grafen (2003) has argued that we can use Fisher's theorem (and the extensions developed by Hamilton and Price) as a licence for what he calls *regulated anthropomorphism*. He explains this as follows:

> Let us ponder this licence a little. Why is a licence needed? Anthropomorphism has been a besetting sin of biologists and others for centuries in understanding organic design. It is essential for a materialist explanation of design to avoid requiring a ghost in the machine. Yet it is also virtually impossible to discuss design without using terms of purpose, so-called intentional terms. To say the eye is *for* seeing is to invoke intention, just as to say that the kidney *processes* waste products, the liver *regulates* blood sugar, or the eye blink is a reflex *to protect* the eye. But a good materialist

needs an excuse for using intentional terms, unavoidable though they are, and a good excuse, a written and logically argued excuse, may be called a licence. (Grafen, 2003, p. 326)

The problem here is that to say that an organ has a function is not to give an intentional explanation. Intentional actions are extensions of natural goal-directed behaviours, viz. those that are or can be done for reasons. An organ cannot give reasons for its activities and it is therefore misguided to claim that when we say that organs have a function we use intentional terms. Furthermore, when a medical doctor explains to his patient suffering angina pectoris that his heart's malfunctioning is caused by tissue necrosis, he does not use an analogy with human (mis) behaviour – as if a malfunctioning heart causing pain is not something real in the living world.

Mathematical theorems are essentially rules for the transformation of empirical propositions concerning, for example, numbers of things, but they do not license us to develop propositions about the goal-directed activities in nature. Fisher's theorem explains the increase of fitness at the population level as the result of selection at the level of the individual, and this theorem can be used to explain the evolution of goal-directed activities. But, as explained above, one should not conflate aetiological and teleological explanations (for Darwin did not replace teleological explanations by evolutionary explanations). How, then, are mathematical equations related to goal-directed phenomena? Recall that these equations are used at two levels. First, they appear in pure mathematical systems and one can 'manipulate' equations by logic and mathematics at this level. For example, we can deduce with the help of mathematical operations Hamilton's rule $rb>c$ from the Price equation (see, for example, Frank, 1998). Accepting deductive proofs at this level is tantamount to insulating the equations from empirical facts. Hence, no theorem is shown to be false by the rejectability of the empirical theories in which it is invoked (a similar point is made by Grafen, 2009). Second, these theorems and equations license the transformation of empirical propositions concerning numbers of offspring. The application to empirical phenomena requires two steps. First, we have to deduce an equation suitable for the phenomenon we are investigating. For example, $rb>c$ becomes

$$\delta W = \tfrac{1}{2}(\delta W_m + \delta W_p) = \tfrac{1}{2}\left(\sum_{i=0} m_i\delta a_i + \sum_{j=0} p_j\delta b_j\right)$$ if it is applied to the

effects of imprinted genes (Haig, 2000). In this equation δa_i and δb_j are the fitness effects of a maternal allele on individual i and a paternal allele on individual j respectively. Mi and pj are the coefficients of relatedness for the maternal and paternal alleles with individual i or j respectively.

There are no differences here between physics and life science. For example, in Newtonian physics, f=ma becomes mg=md^{2}s/dt^2 in the case of the free fall, and this latter equation is used in experiments. Second, an empirical investigation requires a statistical method to calculate relatedness (biologists use programmes such as *Kinship* which estimates relatedness based on data obtained with the help of molecular markers; see Queller and Goodnight, 1989). Yet applying rb>c to empirical phenomena requires also a specification of the inclusive fitness effects of genes. The specification of these latter effects requires a description of the level of biological organization in which we are interested. At this level, differences between physics and life science come to the fore, since we have to answer the question whether we are dealing with chemical reactions, instinctive behaviours, volitional or intentional actions, etc. For we cannot simply assume that organisms are machines and that speaking about purposes in animate nature is merely a form of anthropomorphism. In life science we are dealing with different levels of organization (ranging from autocatalysis to organisms with a mind: see Maynard Smith and Szathmáry, 1995; 1999) and we have to take into account the possible role of goal-directed activities.

4.4.2 Monism and materialism

Distinguishing goal-directed activities as intrinsic parts of nature explains why there is a difference between monism and Williams' materialism (see Williams, 1985). According to (Aristotelian) monism, goals are intrinsic features of the living world and goal-directed activities can be described in terms of what these bring about, for example, the beating of the heart brings about blood circulation (see Hacker, 2007). Morphogenesis results in an organism with organs that has or will evolve its own purposes. And it is the development of volitional and intentional behaviour out of instincts and dispositions that license us to say that humans develop reasoned behaviour (they become responsive to reason), not the immaterial Cartesian ghost in a machine. Materialism, to put matters simply, is Cartesian dualism without the mind. Materialists jettison the immaterial ghost in the machine while retaining the Cartesian idea that organisms are complex machines. They use Paley's term 'contrivance' to argue that the complexity of organisms gives the impression of design. But if teleology is replaced by natural selection, the question arises as to what the argument of design tells us here (see Curio, 1973). If a certain characteristic shows 'signs of design', is this then a proof for the idea that the characteristic is a functional adaptation? According to the alternative view an answer to the question of whether a characteristic has a function and contributes to fitness depends on the results of empirical studies. Malfunctioning internal (e.g. a malfunctioning heart causing pain) and external organs (e.g. leprosy obstructing the use of a hand and hence the development of goal-directed activities) have causal effects, and these may affect survival and reproduction. I shall briefly discuss the example of pregnancy sickness in order to elaborate the differences between the two views.

Pregnancy sickness occurs during the first trimester and is characterized by vomiting, nausea and food aversions. Based on Williams' argument from design, some evolutionary theorists have argued that the symptoms of pregnancy sickness evolved to protect the embryo and foetus against exposure to toxins and pathogens in food consumed by the mother. One argument is that the symptoms occur during embryogenesis and organogenesis (when the embryo/foetus is vulnerable to disturbances). They assume that the placenta becomes after three weeks haemochorial and this corresponds exactly to the onset of the symptoms (Nesse and Williams, 1994, p. 88). Thus the symptoms appear to be designed, for the earlier sources of nourishment are 'much less direct conduits for toxins than is the maternal blood, which nourishes the embryo from the third week until birth' (Profet, 1992, p. 332; see also Flaxman and Sherman, 2000). But does the argument of design prove that the symptoms are indeed functional adaptations? The answer is 'no' and there is a simple reason: *empirical* studies have shown that during the first three months of pregnancy the maternal blood vessels in the placenta are closed by 'plugs', preventing exposure of the embryo to high levels of reactive oxygen species (see, among others, Jauniaux, Gulbis and Burton, 2003). The embryo obtains nutrients during this period from glands in the uterus. Metabolism is during this period essentially anaerobic and the low O_2 tension is maintained for correct cell differentiation. Hence the human placenta is not haemochorial but deciduochorial until the end of the first trimester. It is therefore misguided to suggest that the symptoms of pregnancy show signs of design and are therefore adaptive. We need empirical studies to find out whether the symptoms serve a function and affect survival and reproduction (for an alternative explanation, see Forbes, 2002; Haig, 1993; Smit, 1995b).

The discussion of pregnancy shows why it is highly problematic to use the argument from design for investigating functional adaptations. The example of pregnancy sickness is not an exception, for similar problems arise with respect to other examples (e.g. allergy; compare Nesse and Williams, 1994, chapter 11, with Barnes, Armelagos and Morreale, 1999).

4.4.3 Deleting the mental realm

Williams (1985) advanced the well-known argument that mentalism (based on Cartesian dualism) cannot be empirically tested, since there are no criteria of identity of the mind defined as an immaterial substance. Yet he argued that introducing mental concepts into the biological domain does not add anything to scientific explanations (using efficient causes, i.e. the laws of physical sciences, and natural selection). Hence Williams proposed to delete mental phenomena from biological explanation, for these are, according to Williams, entirely private phenomena. Biology must deal with the publicly demonstrable (Williams, 1992, p. 4).

However, if we delete the mental realm, then it is unclear why a medical doctor uses the expression of sensations by patients for diagnostic purposes. As I have explained in Chapter 2 (section 2.7), there is not a causal relation between the inner and the outer (as Descartes assumed), but an internal relation. The inner is expressed in the outer and we observe the outer manifestations of mental phenomena. Patients express sensations (e.g. pain, fatigue) in non-verbal and linguistic behaviour, and these expressions are used by medical doctors and others as criteria for saying that someone is ill. Moreover, sensations have a location in the body (take the example of angina pectoris), and indicating where a pain is felt may entail relevant information about the possible cause of a disease. Because the expression of sensations through non-verbal and linguistic behaviour does not rest on observing a pain with an inner eye, it is misleading and incorrect to contrast 'publicly demonstrable knowledge' based on medical observations of cells, tissues and organs in the body with the 'private experience' of patients of their sensations. Since sensations are expressed in (linguistic) behaviour, they are simply 'publicly observable'. Williams' misleading contrast is based on Cartesian dualism (see also the discussion by Sullivan, 1986, who neglects Aristotelian insights too and mistakenly argues that medicine excludes 'subjective experience').

The Aristotelian conception of health and disease (see Chapter 2, section 2.8) clarifies why medical doctors use teleological explanations for understanding the function of organs and use the expressions of sensations when investigating patients. If we ignore Aristotelian insights here, this leads to serious misconceptions. The Aristotelian conception also helps us to differentiate between, say, aphasia or a brain tumour and male homosexuality (the stock example against the naturalistic conception of disease). Male homosexuality is not caused by a malfunctioning organ: their putative brain 'disorder' is not comparable to aphasia or to a brain tumour causing disease. Homosexual men are not ill in the sense that their physical health is affected by malfunctioning organs. Moreover, they do not complain of pains like patients with a brain tumour. Hence in the case of homosexuality, there are no arguments for saying that it is a disorder or disease.

Thus instead of deleting the mental realm, as Williams suggested, it is more interesting to ask how the use of mental concepts arose out of primitive expressions of sensations and out of the primitive reactions of our ancestors to the sick (and the dead). Caring about the sick and the way we communicate about mental phenomena evolved during the course of human evolution and raises the question whether the non-verbal and verbal forms of communicative behaviour are explicable in terms of inclusive fitness theory (see further Chapters 5 and 8).

4.5 Purpose, health and welfare

The neo-Aristotelian conception of animate nature teaches us that only living beings can have a purpose. The reason is that living beings have a good, for they can thrive and flourish, decline and decay. Only living beings can be healthy or ill, can be injured, crippled (if they have limbs) and maimed (see Hacker, 2007, chapter 9). We use some of these traits to characterize the stages of their life cycle which have been shaped by natural selection. For if there is genetic variation that affects their likelihood of surviving and reproducing, then organisms with characteristics contributing to reproductive fitness will be selected. Note that things can also be good or bad for a machine (rust and dust, for example), but this is because machines are designed by us to serve a purpose (see Figure 4.1). Hence saying that something is good or bad for a machine is parasitic upon the sense in which things can be good or bad for the human being who uses the machine.

Because the earth is populated by an enormous diversity of creatures, there are interesting differences between their life cycles (see also Chapter 2, section 2.9). For example, plants have a good and have needs (they need water for flourishing), but they do not have desires like animals have. When animals are not capable of exercising their abilities optimally and, hence, cannot satisfy their desires, then this is detrimental to their health. The notion of a good in animals is also linked to the opportunities of successfully engaging in activities that are characteristic of its kind, and, hence, to their welfare.

The notion of a good of a human being is not only a biological notion. Human welfare is associated with the satisfaction of needs and desires, but also with the successful pursuit of projects that human beings adopt in the course of their lives. These projects are both culturally and historically relative. For example, in our society, training in skills, the acquisition of knowledge and the development of intellectual powers are seen as components of the welfare of members of a society, for these enable persons to pursue plans and projects in their lives. Persons can choose these plans and projects more or less wisely with regard to their good. But choosing something can also benefit another person and that is the reason why we distinguish between self- and other-regarding virtues. Notice again that nothing in the life of non-language-using animals corresponds to these (typical human) virtues.

In the context of biomedical science, it is important to note again that the concept of a purpose also applies to an organ. Organs exist for a purpose (have a function) and well-functioning organs contribute to the good of an organism (to their flourishing). However, just as we can distinguish between self- and other regarding virtues, some organs have a function for the organism whereas other organs are beneficial for another organism. For instance, the function of milk glands in women's breasts benefits children, not the women. An interesting observation is that the reproductive organs are clearly for reproduction, for they enable organisms to engage in reproductive activity. Yet animals do not have a notion that a conception results from their reproductive activities and they do not have a desire to reproduce. Copulation in animals is not done for a purpose (it is done in order to satisfy an appetite). Only humans can engage in sex with the aim of producing offspring.

Because humans have goals and pursue plans, flourish or deteriorate, are healthy or sick, their good may be protected (preventing deterioration), cured (making their condition better) or promoted (augmenting their good). Their health is associated with survival and is therefore a component of their fitness. But, as I have explained in Chapter 3 (section 3.2), health and survival are not distinct parts of the individual's fitness. This simplification is innocuous as long as evolutionary theorists realize that Darwin did not develop his theory as an extension of the Cartesian conception of living beings.

4.6 Teleology and group selection

A theoretical reason why evolutionary biologists ignore teleology may be group selection. For saying that a well-functioning organ is good for the *whole* organism may mistakenly be taken as a form of group selection (it is

mistaken, for it means that a well-functioning organ is partially constitutive for the health of the organism, and this is not an evolutionary
explanation). However, the older group-selection hypothesis has been
advanced by biologists and philosophers in the past (see Chapter 3, section
3.4). For example, Megone (1998, 2000) has used the older hypothesis in
order to justify the use of the phrase 'that a well-functioning organ is
good for the something'. He argued, just as I have done, that the function
of organs is subservient to the good of the whole being, just as the
function of parts of a clock is subservient to the function of the whole
clock, i.e. indicating the time. But if we cannot refer to a Designer, what,
he asks, is then the basis for using teleological explanations? Megone argues
that the claim that the activity of an organ constitutes a function requires
that the activity is good from some *perspective*. Notice that this requirement does not arise if we simply distinguish the function of an organ (the
activity of the organ contributing to the health of the organism) from
the function of a part of a machine designed by us. Megone argues that
'the heart's function must be a goal, good from some perspective, and
pumping the blood can be seen as achieving a goal if that activity contributes to the persistence of the species, and that in turn is, from some
perspective, good' (Megone, 2000, p. 57). The perspective subsequently
advanced by Megone is that goal-directed activities contribute to a stable,
ordered, persisting ecosystem and that this is better than a chaotic, degen-
erating ecological environment. This seems to me misguided since Megone
invokes the older group-selection hypothesis. Evolutionary biologists may
rightly object to this explanation. They have argued that inclusive fitness
theory offers more promising explanations. For example, one can argue
that a heart executes its function since its activity contributes to the
inclusive fitness of the genes residing in its cells; for copies of these genes
are also present in the germ cells. Hence when organisms were constituted,
the major way for genes to replicate was for the whole genome to get
replicated, i.e. for the organism to survive and reproduce. This explains
the emergence of the genome as a functional collective: all the alleles
have the same interest in producing an organism with organs. It also explains
the suppression of selfish alleles whenever there is a functional collective (see
among others Frank, 2003; Michod *et al.*, 2006; Okasha, 2006).

It is a challenge for evolutionary biologists to explain why during
evolutionary transitions functional collectives arose that are to a large
extent, as it were, immune to selfish alleles. Yet many empirical studies
have shown why we have to add the phrase 'to a large extent'. For instance,
the evolution of imprinted genes demonstrates that there are opportunities

for intragenomic conflicts (Haig, 2002; Smit, 2009). But there are also examples showing that selfish alleles evolved of which we, as rational beings, say that they are probably a dead end. A well-known example is the facial cancer occurring in the Tasmanian devil. The cancerous cells arose as the result of a chromosomal mutation and are transmitted from one individual to the other by biting or by shared feeding. The cells are not recognized by the immune system as foreign since there is not much diversity in the immune system (i.e. the MHC system) of the devil since the population went through a bottleneck some decades ago. This cancer spreads rapidly in the population and there is evidence that the cancerous cells evolve towards a more virulent type. Studies have shown that the cancer cell underwent at least ten mutations during the period that this disease was studied (McCallum and Jones, 2006). As a result of the disease the population will probably go extinct (although there is evidence that there are resistant types in the population). This example shows that increase in cooperation does not always mean reduced conflict and there are other highly cooperative groups that have very high levels of conflict (Queller and Strassmann, 2009; Strassman and Queller, 2010). Yet it also shows that returning to selfish behaviour at the level of cells is difficult, just as it is difficult for humans to return to asexual reproduction after the evolution of imprinted genes.

4.7 Conclusion

Teleological explanations are a distinct type of explanation. They are not causal explanations and it is therefore unsurprising that biologists, living in an era in which causal explanations are the norm, are reluctant to admit that they use them for understanding phenomena. This is nicely captured in a saying of Haldane (cited in Lennox, 1992): teleology is like a mistress to a biologist; he cannot live without her but is unwilling to be seen with her in public. Yet teleological explanations are not indicative of defective knowledge and it is still a challenge for evolutionary biologists to explain the evolutionary origin and maintenance of the several goal-directed phenomena in animate nature.

I have suggested that it is greatly preferable to study goal-directed phenomena as intrinsic parts of nature rather than as phenomena explicable by analogy with human intentional behaviour and/or with the help of Williams' argument from design. Distinguishing goal-directed phenomena as intrinsic parts helps us to resolve conceptual problems and enables us to develop more interesting hypotheses. Replacing teleology by natural selection, by contrast, conflates the conceptual and the empirical, leading to misinterpretations of empirical phenomena. Hence further investigations of the role of teleology in our conceptual network will clarify what belongs to the conceptual net and what to the empirical fish that we catch with it. These investigations will also contribute to the further elaboration of the field of evolutionary medicine and psychology.

Darwin's rehabilitation of teleological explanation can easily be integrated within the framework of modern evolutionary theory. I have distinguished three types of goal-directed phenomena and have argued that the formation of intentions occurs only in human children. For only a language-using creature learns to announce an action ('I am going to V'), learns to give reasons for its actions and, hence, learns what count as adequate reasons. And because only humans understand that certain actions are desirable or obligatory, they acquire a moral sense (see further Chapter 8). Aristotelians, in contrast to Cartesians who emphasize self-consciousness as the mark of the human mind, emphasize that both animals and humans display goal-directed behaviour, but that only humans form intentions. I shall elaborate this difference between the Aristotelian and Cartesian in the next chapter, and its consequences for studying conflicts in the mind.

CHAPTER 5

Dualism, monism and evolutionary psychology

5.1 Introduction

According to evolutionary theorists, human behaviour can be resolved into two sources. The first source is hereditary: through decoding of information in the human genome, structures and mechanisms develop in our bodies and brains that influence the development of behaviour. These genes have been selected during our evolutionary past, since they contributed to our reproductive success through their phenotypic effects. The second source is culture: humans acquire knowledge through internalization of ideas and use these ideas as a guiding principle for their actions. Some of these ideas are present in our culture because, like genes, they have proven their value in the long term. Yet cultural evolution evolves much faster than genetic evolution, since certain ideas may be replaced by other ideas within a single generation.

Genes and ideas influence behaviour through different 'means' because they have different 'goals'. According to evolutionary theorists, genes influence behaviour through instincts and ideas affect our behaviour if they are stored in our memory. Haig explains this as follows:

> Instinct summarises the wisdom of past natural selection and recommends actions that have worked before under similar circumstances. Culture also summarises wisdom from the past and can respond much faster than gene sequence to environmental change, but, from a gene's eye view, has the disadvantage of evolving by rules that need not promote genetic fitness. (2006, p. 11)

Besides the guiding operations of instincts and ideas, humans have an intellect that enables them to act in unique situations. But our intellect 'may lack the historical judgment of either instinct or culture' (Haig, 2006, p. 11). If, for instance, someone has sexual intercourse just to enjoy himself and uses (for good reasons) a condom, then these rational considerations interfere with the goals of instincts: the dispersion of genes. And if he is a

Catholic priest and allows his celibate life to be disturbed by sexual longings, then his behaviour is, according to Haig, a violation of an idea. Although these examples make clear that the 'actions' of the intellect may interfere with the long-term goals of genes and ideas, one cannot infer that humans always behave irrespective of their instincts and ideas. The opposite possibilities occur too: some people behave as if they are slaves to their instincts, while others behave according to the letter and spirit of certain ideas. The possible conflicts between instincts, ideas and the intellect are traditionally explained by reference to the fact that human beings are rational creatures with instincts and a conscience.

The variable nature of the influence of genes, ideas and the intellect is an interesting finding for evolutionary theorists: if there is variation, there might be selection. Accordingly, the theory of evolution may explain possible outcomes. Evolutionary theorists are particularly interested in how selection has moulded conflicts in the mind. If there have been conflicts between genes, ideas and the intellect, then which evolutionary forces have moulded possible outcomes? Do these forces tell us something about the architecture of the mind? What are the possible consequences of selection pressures on, for instance, the strength of our will? These problems are studied by evolutionary theorists and some interesting hypotheses have already been proposed. These hypotheses are also answers to problems traditionally studied by psychologists and philosophers. In this chapter I will analyse the conceptual framework used by evolutionary theorists when they study conflicts in the mind. I will focus on ideas they have taken from the psychology of William James (1950 [1890]). First, I will discuss the idea that instinct is more important for understanding human behaviour than some social scientists have thought in the past. Unlike social scientists, who argue that human beings are rational creatures with few instincts, evolutionary theorists emphasize, just as James did (1950 [1890], vol. 2, p. 393), that humans are special because they have many instincts. Since the human mind entails, according to evolutionary theorists, a lot of domain-specific instincts (also called psychological mechanisms or modules), we are able to behave flexibly in the most diverse situations. Second, I will discuss the idea that there are conflicts in the mind between genes and ideas (or memes, as Dawkins has called these cultural units of information). Like James (1950 [1890], vol. 2, p. 526), evolutionary theorists insist that instances of antagonistic thoughts manifest conflicting 'agents' in the mind. They have advanced the hypothesis that natural selection has moulded the outcome of such conflicts.

I shall argue that the conceptual framework based on these ideas of James is incoherent and inappropriate for studying human and animal behaviour. For evolutionary theorists inherit via James' ideas an unhappy marriage between Darwin's theory and a (crypto-) Cartesian conception of the mind (and the brain). I will argue that the alternative, Aristotelian concept of the mind is a better starting point for developing a coherent framework for studying animal and human behaviour. This framework enables us to develop a clearer picture of conflicts in the mind.

I start with a discussion of the main differences between the Cartesian and Aristotelian conceptions of the mind and analyse how these conceptions are reflected in ideas about the mental capacities of animals and humans. Although evolutionary theorists repudiate Cartesian dualism, they do not endorse Aristotle's monism. I will argue that their theory may be described as a crypto-Cartesian conception of the brain or mind. Next, I discuss why the flexibility of human behaviour evolved not because humans have more instincts than animals, but because humans started to display intentional behaviour on becoming language-using creatures. Although animals display goal-directed behaviour, they do not develop intentional behaviour since they lack a language. Since evolutionary theorists have argued that there are conflicts between instincts and intentions in the human mind, I will discuss how intentional behaviour develops out of goal-directed behaviour (which is also present in animals). Finally, I analyse the evolutionary explanations of possible conflicts between genes and ideas. I discuss why (crypto-) Cartesian ideas on intrapsychic conflicts are mistaken and why an Aristotelian framework extended with Darwin's theory is a better starting point for investigations.

5.2 Monism and dualism

The opposition between monism and dualism is a central theme in Western philosophy (discussed at length in Hacker, 2007). According to the monism of Aristotle, a characteristic of living beings is that these consist of a substance with a form. As a biologist, Aristotle made a distinction between things and living creatures with the principle that only living beings have a *psuchē* (often translated as soul). It is important to keep in mind that the soul is a biological principle and that the possession of a soul is not a characteristic of human beings alone. According to Aristotle, plants and animals have a *psuchē* too.

The soul is not distinct from the body, as in the dualism of Descartes. It is not a part of the body or a part of the mind. There are no interactions

Table 5.1: The Aristotelian (A) conception of the mind.

A:	The mind is not a substance and is not an agent.
A:	It is not a part of a being and does stand in a causal relation to the body.
A:	To possess a mind can be described in terms of the (manifestations of) powers.
A:	The powers typical for the human mind are linked to responsiveness to reasons.

between the soul and the body, as Descartes assumed. The questions as to whether the soul belongs to the body or is identical to the body cannot be answered, according to Aristotle, as such questions presuppose an incoherent dualism. The *psuchē* is the form of the living substance and observable in the abilities of living beings (see Table 5.1). Plants can reproduce, grow and metabolize nutrients. They have what Aristotle called a vegetative soul. However, plants are not self-moving creatures and do not use senses as animals do. Animals use their senses to search for food or a mate. For that reason, Aristotle thought that animals have a sensitive soul that plants lack. Yet if we say that animals can observe things in their surroundings or are able to focus their attention on something, we do not apply these psychological predicates to their soul. Their behaviour, as Wittgenstein later remarked, is the criterion for applying these predicates to them. Since the *psuchē* is visible in the behavioural manifestations of animals' capacities, Aristotle states in *De Anima* that:

> to say that the *psuchē* is angry is as if one were to say that the *psuchē* weaves or builds. For it is surely better not to say that the *psuchē* pities, learns, or thinks, but that the man does these things with his *psuchē*. (2002, 408b12–408b15)

The same insight was formulated by Wittgenstein (2009 [1953]; see also Chapter 6 below).

Aristotle used the concept of *psuchē* to sort out the differences between plants, animals and humans. He stated that animals have a vegetative and sensitive soul, but that only humans have a rational soul (the intellect and the will). Since we know now that humans have gone through an evolutionary history, we can more precisely formulate why only humans are rational beings: only humans have developed a language. As language users, human beings are responsive to reasons and, hence, may be motivated to act for and by reasons. This is a major difference between humans and animals, since animals cannot respond to reasons or engage in reasoning from premises to conclusion. However, this does not mean that animals are complex machines, as Descartes thought. Animals have goals and display goal-directed behaviour. They can choose, since they have preferences and avoid situations or ignore food that they dislike. Yet animals cannot make decisions that are the upshot of practical reasoning that involves considerations and weighing of reasons. Consequently a crucial aspect of an Aristotelian framework extended with Darwin's theory is not the evolution of *Homo sapiens*, but the evolution of *Homo loquens*, since the evolution of language enabled humans to become beings with a rational *psuchē* responsive to reasons.

In his dualist philosophy Descartes offers a different conception of the human mind. Put in Aristotelian terminology, Descartes argues that animals do not have a rational and a sensitive soul. Animals have only a vegetative soul and this soul is, according to Descartes, fully understandable in terms of the models of mechanics. These ideas result in the well-known conclusion that animals are complex machines without a mind. Humans are machines with a mind since only humans have the capacity to think (there exists a separate thinking substance present only in humans). The capacity to think is not, as in the philosophy of Aristotle, the form of a substance that is visible in behavioural manifestations, but an essential property of an immortal immaterial substance. This substance is describable in terms of the conscious experiences someone has. These experiences are private and not essentially manifest in behaviour, since these experiences are present in someone's mind (and, hence, are separated from what happens in the body). They are an indubitable source of knowledge. Since what Aristotle called the sensitive soul also belongs to the mind, perceptions and sensations (like pain) belong to this substance. Descartes thought that animals do not suffer pain, because all their behaviour is understandable in terms of mechanical models. Descartes assumed that if someone has pain, he feels this pain in his mind *as if* the pain was present in the body. And perceptual experiences properly speaking (seeing a coloured object, hearing a sound, etc.) are modifications of the mind caused by (although logically independent of) the impact of particles upon nerve endings (see Table 5.2). But animal sensation, perception and appetite are fully explicable as non-conscious reflex actions.

Dualism raises questions (such as, where is the mind located if the mind can interact with the body?) that are unanswerable according to the proponents of monism. There are no solutions to these questions, since the conceptual picture of the mind and the body given by dualists is incoherent. This critique of dualism explains why this philosophy has given rise to popular discussions but not to scientific research. With the emergence of the neurosciences the dualism of Descartes was criticized and

Table 5.2: The Cartesian (C) and crypto-Cartesian (CC) conception of the mind.

C:	The mind is an immaterial substance distinct from the body.
CC:	The brain is the material organ of the mind.
C:	The mind is the subject of mental or psychological concepts.
CC:	The brain is the subject of mental or psychological concepts.
C:	The mind is an active agent and interacts with the body.
CC:	The brain is an active agent and stands in two-way interaction with the body.
C:	The contents of the mind are private.
CC:	The essence of the material mind is the qualitative character of private experiences.

replaced by alternative materialistic views of the mind (discussed in Bennett and Hacker, 2003). A striking feature of this shift from dualism to materialism is that the Cartesian framework was not abandoned, although the Cartesian distinctions are now made in terms of the brain (sometimes called the mind/brain). So in the materialistic view of the brain, the brain is still an agent that can do things and is in this sense similar to the agent mind in Descartes' conception. A well-known example of this materialistic view is the idea of the mind/brain as a computer. In this view the brain is the hardware, our mind the software. Accordingly the relationship between mind and brain seems easier to describe as there are no interactions between two different substances. However, perception is explained by assuming that the brain is an information-processing device. When signals from the senses are transmitted by afferent neurons to the brain, then the resultant state of the brain is said to be the perception of something. So it is the brain, not the individual, that perceives. Likewise, if someone wants something or has the intention to do something, then there is a state of the brain that, through the efferent neurons, moves the body. Again it is the brain, not the individual, that is the originator of the intentional movement. Within this framework, self-consciousness is a self-scanning mechanism and so the knowledge we have of ourselves is regarded as an active mechanism in the brain. Owing to the similarities between the original Cartesian view of the mind and this view of the brain, the materialistic view of the brain is aptly described as the crypto-Cartesian view of the brain (see Table 5.2). Therefore, although scientists and philosophers say that they have taken formal leave of the Cartesian conception, the main structure of the Cartesian framework remains intact. I will argue that the ideas of evolutionary theorists taken from the psychology of James fit within the crypto-Cartesian conception of the mind/brain.

5.3 Instincts and behavioural flexibility

Instincts are conceptualized by evolutionary theorists as (during the course of evolution) selected mechanisms that are present in the brains of humans and other animals. If they are triggered by stimuli, they generate behaviour. A good example is the behaviour of a turkey hen when she is rearing offspring. During this period she will attack every approaching, moving object that does not produce the sound of hatchlings. That this behavioural response may be seen as a mechanism in her brain became clear when scientists studied the example of a hen that attacked her own offspring: she was deaf (Schleidt, Schleidt and Magg, 1960). So in this case there appears to be a mechanism in the brain that, in normal cases, is triggered by specific sounds and results in behaviour that protects hatchlings against predators. Hence this mechanism is responsible for a behavioural response that increases the inclusive fitness (for other examples of instinctive behaviour in animals, see the classic study of Tinbergen, 1989 [1951]).

The behaviour displayed by turkey hens when they rear offspring is an extreme example. There are other examples which show that animals are able to behave flexibly (see, e.g. Bonner, 1980). For example, a female lion becomes hungry because of internal, hormonal signals. In order to appease her hunger, she will have to catch a prey. Therefore she has to explore the environment in order to spot a suitable prey, for example, a zebra. If she has spotted one, she can, perhaps in cooperation with other females, try to attack the zebra. But if it turns out that the zebra spotted is not easily outwitted, then the lion has to change her plans and will opt for an attack on another zebra. Based on the sequence of the different actions, the fact that there are options and the fact that lions make choices and are able to readjust their goals, we can say the females behave flexibly, since they adapt themselves to unique, local situations. While the behaviour of the turkey hen is an example of unconscious, instinctive behaviour, since it is a species-specific response to an environmental stimulus, the behaviour of a female lion is an example of conscious, goal-directed behaviour.

These examples raise the question as to why there is variation in the flexibility of behaviour. Two explanations are proposed. The first explanation makes clear why there is variation in flexibility within one species. Evolutionary theorists expect less flexible behaviour to evolve in fitness-relevant situations that recur in the life cycle of a species (like parent–offspring interactions or the mating season). If these situations recur and if behaviour displayed by individuals in these situations has fitness consequences, then there will be selection in favour of genetic variations that increase the fitness in these situations. The rearing behaviour of the turkey hen started as a rare genetic variation but has been selected because the allele responsible for the variation led to an increase in the inclusive fitness. This behavioural response therefore invaded the population, superseded older forms of responses and gradually became a species-specific response. This is a plausible scenario for the evolutionary origin of the discriminatory response of the females: turkeys that were not able to discriminate between the sounds of hatchlings and other 'moving objects' were put at a disadvantage. However, it does not follow that the behavioural response is 'genetically determined'. Evolutionary theorists recognize the possibility that learning processes may also be important during the development of the behaviour. However, since the behaviour of female turkeys is subject to strong selection, they expect that genes play a role in the development of the response. That explains why this is an example of rather rigid behaviour.

The second explanation is that there are differences in flexibility between species. Species that have more instincts are able to behave more flexibly. Tooby and Cosmides (1992) state that 'to behave flexibly, humans must have more "instincts" than other animals, not fewer' (p. 93; see also Buss, 1999, chapter 2; Pinker, 1994, p. 20). For an explanation of this statement I will start again with the example of the female turkey's behaviour. These females have an instinctive mechanism in their brain that enables them to *discriminate* between different sounds. However, this ability to discriminate is not based on knowledge. They can respond differently to various sounds, but there is no basis for stating that the females *classify* certain sounds as dangerous. Yet suppose there is a new species with a new instinct: the ability to classify sounds. We can state that the members of this new species know what they do. If an individual of this species classifies a sound as dangerous (e.g. as the sound of a predator), then this will be a reason for this individual to become anxious. Should this individual have offspring, it will take care of these. However, if this individual discovers that it has made a mistake (the sound classified as dangerous turns out not to be the sound of a predator), then the anxiety of the individual will rapidly disappear, since the reason for the anxiety no longer holds. This individual realizes that it has mistakenly classified a sound as dangerous. Such an individual can change its behaviour very quickly, because it can make mistakes *and* is able to recognize these. This is clearly different from an individual that can only discriminate between sounds, because the latter individual can only exhibit different responses to stimuli and, therefore, cannot recognize mistakes. Consequently, this example may suit evolutionary theorists since it illustrates why a species with a new instinct may behave more flexibly. There are, however, some problems in this argument. First, there is only one species whose members display behaviour that demonstrates this alleged instinct: the human species. This is because the ability to classify has a normative dimension (see Baker and Hacker, 2009). If someone mistakenly classifies something as dangerous, then he can explain why he at first thought that there was a danger and why the perception was a reason for his anxious behaviour. This is only possible in the case of a creature displaying intentional behaviour: that is, one that can give reasons for its behaviour (and animals cannot give reasons). The evolution of the ability to make distinctions based on classifications is, therefore, not simply an extension of a new instinct. And neither is it a mechanism in the brain, since the discovery or recognition of a mistake is not a neurophysiological phenomenon. Second, this argument underestimates a possible role of (what Aristotle called) the sensitive soul in the evolution of flexibility. Since non-human animals have sensitive souls and, hence, can use their senses, they can learn by trial and error. Animals cannot make mistakes in the normative sense, but they can discern an error. At least some animals learn to associate a given sound with a dangerous predator. Therefore we can say that they recognize the sound of a predator, respond by avoidance behaviour and are able to recognize an error in the sense of ceasing to behave in a given way in response to perceiving that things are not as they previously took them to be.

These problems recur in interpretations of empirical data. Evolutionary theorists have argued that genes will affect our choices and preferences through evoking emotions in fitness-relevant situations (see Tooby and Cosmides, 2005). Emotions are in humans, as Haig (2006) has put it, the carrots and sticks of genes. Hence emotions are seen as mechanisms in our mind/brain. Although emotions are not instincts that 'force' us to act in certain ways, they increase, according to evolutionary theorists, the chance that we will perform certain acts or develop specific preferences. For instance, a study in Israel demonstrated that during the possible threat of an attack by missiles in the first Gulf war, people increasingly contacted family members and temporarily 'neglected' friends and colleagues (Shavit, Fischer and Koresh, 1994). According to evolutionary theorists, this change in their behaviour can be explained with the help of kin selection, since the emotion of fear in this particular situation triggers a preference for kin and an accompanying mechanism in the brain redirects our attention towards family members. This explanation of the change in behaviour resembles the change in behaviour that occurs in turkey hens when they are rearing hatchlings. In both cases there appears to be a stimulus that triggers a mechanism in the brain that generates behaviour. Kin selection explains why these species-specific mechanisms have been selected. There are, however, differences in behavioural flexibility. But these are not explicable by referring to the fact the Israelis have more instincts. The relevant conceptual difference is that there is an object of fear involved (see Kenny, 1963, chapter 2). The change in the behaviour of the Israelis may be seen as a response to the threat of a missile attack. However, this change is, at least in part, described by reference to the object of their emotions (what they were afraid of). The scientists studied not only the change in their physiological response (like the increase in the heart rate or the display of an anxious facial expression), but also the worry or concern about family members. The change in my behaviour when my daughter is in a danger-ous situation is visible not only in the change of my autonomous nervous system, but also in what I will do: how many phone calls I make; whether I arrange an air-raid shelter for my daughter, etc. For the same reasons, we determine someone's fear of heights not by observing his physiological response, but by noticing the efforts he makes to avoid heights. Knowledge is important for how someone can deal with the object of an emotion. Suppose that the Israelis discover the fear of an attack is unwarranted (since the threat of the missile attack was a faulty assessment), then the emotion will cease since their fear is no longer warranted. The change in the relevant properties of the object explains the change in behaviour. Hence we refer to the object of the emotion in order to explain the flexibility in their behaviour, not to a mechanism in their mind/brain.

Animals display simple or primitive emotional behaviour, and the object of an emotion may explain their actions too. If an animal is threatened, then it is relevant to know what the threatening object is and how the animal responds to it. Suppose that a chimpanzee, which has a low rank within the group, notices the crack of a twig. Then he may become anxious because of the possible approach of the alpha male. This is similar to the anxious response of someone lying in bed on hearing a noise downstairs: it may be the sign of a burglar (but the noise may also be produced by the family's dog). In both cases the noise causes the anxiety, yet the object of the emotion makes clear what they are frightened of (alpha male or burglar). The behaviour displayed by the chimpanzee shows that he is afraid of the alpha male since he is alert, and his moves show that he is fearful of a possible attack by the alpha male.

There are, of course, differences between the emotional behaviour displayed by animals and humans. If animals fear something, flee or avoid situations, then concrete situations are involved. If a cat sees the basket and disappears, it may do so in order to avoid a visit to the veterinarian, because the cat has made an association between the basket and an unpleasant visit. Yet humans may also feel fear due to a visit to the dentist next week, because they know now what to expect then. There is, however, nothing in the behaviour of cats that shows that they are now afraid because they know that they have to visit the veterinarian next week. Only a language-using creature can be concerned now about a determinate event at some determinate time in the future (e.g. next Wednesday or next Easter). Therefore complex emotions are typical for humans. A parent may be proud of a child and hope that he or she will get a good mark for the exam next week. The love someone feels for another may express itself in daydreaming about a romantic dinner during the weekend. Or, if you prefer a less romantic example, someone may be eagerly looking forward to a visit to a peepshow on a Saturday night.

Animals are not machines without a mind, but their emotional behaviour is less rich than human emotional behaviour because human emotions involve long-term desires, fantasies and wishes. Like animals, humans have a memory, but unlike other animals, humans have a cognitive and doxastic past and a future (only humans can recollect). This is not surprising since only humans use a language. By studying the role of language in the development of behaviour, we can, according to Aristotelians, get a better idea of the differences and similarities between human and animal behaviour.

5.4 The role of language

In contrast to the (crypto-) Cartesian conception of the mind/brain, Aristotelians have argued that the flexibility in human behaviour arises when children become language-using creatures. With the help of two examples I will first describe how the flexibility in human behaviour arises as an extension of natural, instinctive behaviour that is also present in animals. The *locus classicus* where this has been substantiated is the *Philosophical investigations* of Wittgenstein (2009 [1953]).

5.4.1 Sensations and emotions

Sensations and emotions have expression in the behaviour of animals and humans. If children have pain, they will display pain behaviour like crying; if they are angry or anxious, they will display species-characteristic emotional behaviour, like a scared facial expression. Since animals display sensations and emotions in their behaviour too, we can apply these psychological predicates to them as well. By comparing the form or structure of the different species-specific facial expressions, scientists are able to reconstruct the phylogenetic relationships between different species (see van Hooff, 1972 for an example).

Sensational and emotional behaviour transforms as children learn to use a language (Malcolm, 1982). Their behaviour is then extended with and replaced by learnt linguistic expressions like 'I have pain' or 'I am angry.' These linguistic utterances are not based on introspection, in the sense that a child first observes his or her sensations or emotions with an inner eye, and thereby learns that he or she is in pain or is angry. Wittgenstein (2009 [1953]) has argued that these utterances are expressions of emotions and sensations themselves and may be seen as new forms of pain and emotional behaviour. These are learnt extensions of natural behaviour and create new possibilities for communication about sensations and emotions. If a child learns to use a language, it will be able to answer the question where a certain pain is located (since sensations, but not emotions, have a location in the body), and may give a possible cause of the pain ('I have fallen'). If children are angry, they can, as language-using beings, answer the question why they are angry and what the object of their anger is. Whether crying is an expression of pain or fear is, in certain circumstances, no longer confusing: if a child has pain, it can explain where the pain is located, and if it is anxious, it will refer to an object of a fear ('A scary dog'). The emotional life of children becomes still more complex when children develop an inner life (Hacker, 1990b). An inner life arises when, through playing as-if games, children learn to pretend. If a baby crows with pleasure, there is no reason to suppose that the child has a rich inner life, since the child has not yet learnt to conceal its emotions. Older children learn the difference between revealing and concealing emotions, they learn the difference between being sincere or insincere, or learn to deceive others or to deceive themselves. Therefore, although both animals and humans express their sensations and emotions in their natural behaviour, it is only in the human species that this natural life is transformed into a complex, partly rational, emotional life interspersed with cognitions. During a child's development, the natural expressions of emotions become less important for understanding their emotional life. These expressions become the residues in their behaviour that show the continuity of child and adult, and the kinship between animals and man (see also Hampshire, 1961; 1965).

5.4.2   Intentions

First-person present-tense utterances of sensations and emotions are extensions of species-characteristic behaviour. At face value, there does not seem to be a precursor for intentions, since expressions of intention, both in advance of acting and after acting, appear to be purely linguistic. For example, if I say that 'I have the intention (expectation, belief, thought) to take the train to Rotterdam tomorrow', then I am explaining my intentions or plans. A listener then knows what I will do tomorrow, although I may, of course, modify my plans because of new information or alternative considerations. Animals cannot express intentions, since they are not language-using creatures. So must we therefore conclude that there is no precursor of intentional behaviour in natural behaviour? There is a precursor of intentional behaviour and that is goal-directed behaviour (Hacker, 2007, chapter 7; Kenny, 1989, chapter 3; Wittgenstein, 2009 [1953], par. 647). When animals and children grow older and start to explore their environment, they will encounter attractive or repulsive objects which draw their attention. They will act in pursuit of what they want, start to explore and to experiment with objects or use them as a toy. There are no differences here between animals and children, since both display goal-orientated behaviour, which we can infer on the basis of the sequence of their behaviours. If we see, for instance, a young cat stalking a bird, it is the sequence of its movements that allow us to infer what it is after. Yet although goal-orientated behaviour is a precursor of intentional behaviour in the human species, animals do not develop intentions. Children learn intentional behaviour when they start answering the what- and why-questions posed by their parents. As a child learns a language, he or she initially learns some names of objects ('car', 'doll') and next utterances like 'want car' or 'I want car.' If children start to play with these objects, then parents may ask them what they are doing, or why they are doing something: 'What do you want to achieve?' Or, 'Why are you doing this?' By answering these questions, a child learns to describe her own behaviour and to justify it. Still later a child learns to make announcements ('I am going to do such-and-such, because I want . . .'). He learns to develop plans. If a child is able to give reasons for his plans, then he develops primitive forms of intentional behaviour, which at the same time form the beginning of his rational life.

These two examples demonstrate how children, through learning a language, are becoming members of a unique species in the animal kingdom. Building on characteristic, natural behaviour and goal-orientated behaviour, which is also present in animals, they develop a complex emotional and intentional, rational life. The distinction made above between animal and human behaviour does not coincide with the Cartesian distinction between animals as 'machines without a mind' and humans as 'machines with a mind' (Smit, 1995a; 2007). According to the Aristotelian view, animals have all kinds of abilities that enable them to respond on a moment-by-moment basis to changes in their environment. Animals can observe, learn, have a memory, are conscious (but not self-conscious), have purposes, etc. Mastering a language, however, makes the life of a human being richer, because humans can develop plans for the future or explain their behaviour by referring to events in the past. When a dog walks to its master with its collar in its mouth, it can perhaps express the wish that it wants to go for a walk. However, the dog cannot express the wish that it wants to make a walk to the butcher next Wednesday, because it is not a language-using being that can express this wish in the future tense. If a lady says that she knows for sure that her pet wants a bone for Christmas, we would not know why she utters a nonsensical sentence since she misuses the rules for the application of psychological concepts to creatures. What behaviour of the dog induced her to say this? Since there can in principle be no observable evidence, no criteria for her assertion, we do not know what it means for a dog to want a bone next Christmas. Of course, the lady may plead for a relaxation of the criteria for the use of psychological predicates in the case of animals. If, for example, a group of lions attacks a zebra, is that not an example of a planned, intentional act? Again, lions do have purposes but do not develop intentions because they cannot give reasons for their primitive plans. Nevertheless we can narrow down the concept of 'intention' so that it is applicable to lions, or small children for that matter, too. However, this is a decision we make not on scientific grounds, but on conceptual grounds. And if we do so, it does not follow that we can call lions up to a tribunal because of their 'intentional killings' of zebras, or, in the case of males, because they have killed the young when they took over a pride (as if the male has committed a crime). If we apply the concept of intention to animals, then we call their intentional behaviour simple or primitive, since the primary application of this concept is to a language-using being.

5.5   Intrapsychic conflicts

According to evolutionary theorists, possible sources of conflicts in the mind are conflicts between genes and ideas. If someone is willing to die for an idea, then his behaviour results in a reduction of his reproductive fitness. Or, to discuss a less extreme example, if a woman is an extreme weight-watcher, her menstrual cycle will be suppressed and this shortens her reproductive period. In these cases there is a potential conflict between genes and ideas since behaviour guided by ideas reduces the reproductive fitness. Evolutionary theorists investigate whether and how natural selection may have moulded possible outcomes of these conflicts. They wonder whether it has been possible for genes to counteract the effects of ideas through developing mechanisms in the brain that diminish fitness-decreasing effects of ideas in conflict situations. (Some scientists think that ideas are also able to adapt themselves and develop characteristics that increase their spread in populations; see Dawkins, 1976, chapter 11.) Is it possible for genes to counteract rational decisions if there are conflicts between genes and ideas? If humans were not able to 'control the effects of genes' fully, and if conflicts between genes and ideas took place in recurrent situations (i.e. situations that always occur during specific stages of a life cycle), then in theory this is possible. What, then, are the possible expressions and outcomes of these conflicts? Evolutionary theorists, following James (1950 [1890], vol. 2, p. 526), have argued that in situations where there are conflicts between genes and ideas, decisions that are disadvantageous for genetic fitness will require mental effort (Haig, 2006; see also Dennett, 1991, chapter 12). They have proposed two possible mechanisms.

First, genes may generate impulses that decrease the probability of someone taking rational decisions that are disadvantageous for genetic fitness. Evolutionary theorists assume that the impulses (generated by genes) act autonomously; that is, independently of what someone wants. However, what exactly do we mean if we say that someone has uncontrollable impulses if he wants to make a rational decision? That he cannot resist the temptation to do something, even though this is in conflict with rational considerations? If an impulse is indeed irresistible, then it causes him to act and makes him do what he then does. In such cases someone is in the grip of an impulse. Yet if someone cannot control impulses, then his behaviour is no longer intentional because he cannot choose between different alternatives. His behaviour is an example of an unnatural behaviour that occurs when someone is, for instance, addicted. So if someone ought to behave intentionally but cannot control his impulses, then his behaviour is no longer an example of intentional behaviour. His behaviour may be the result of psychopathology. For instance, some people scarcely display empathic behaviour in situations where most people experience empathy. In some cases they have fantasies of raping someone and use instrumental aggression. If these people violate norms, then there are good reasons to study (the role of genes in) the development of their behaviour. For instance, in some cases of psychopathy there is evidence that the deviant developmental trajectory psychopaths go through has an onset when they are about 4 years old. The extreme psychopath does not appear to develop an emotional sensitivity for distress signals displayed by another person (Blair, Mitchell and Blair, 2005). If family and twin studies demonstrate that genes are important for this deviant trajectory (see, e.g. Caspi et al., 2002; Meyer-Lindenberg et al., 2006; Viding, et al., 2008), then it makes sense to consider whether the prevalence of these genes is understandable in terms of evolutionary processes (like frequency-dependent selection).

Second, selected mechanisms in the brain may, according to evolutionary theorists, explain why some decisions that are in conflict with genetic fitness require good reasons. For if there is variation with respect to the vigour with which decisions are taken that conflict with genetic fitness, then there may have been selection against 'easy decisions' because people who made decisions with greater ease left fewer descendants since they passed over the interests of the genes and, hence, had low reproductive success (Haig, 2006). If there are fewer effects of a rational decision on reproductive fitness, then an individual may be less motivated to make the decision. Haig has argued that natural selection is able to mould the strength of our muscles and the strength of our will. However, the strength of the will is not comparable to the strength of our muscles. If someone wants to act, that means that he can choose, has different options and knows which goals will be obtained by choosing an option. He can give reasons for his choice. Of course, someone may hesitate and doubt whether a certain option is the best one to choose. He has not made up his mind and can describe his indecisions as a conflict. However, evolutionary theorists assume that there is a mechanism (like a voice in someone's brain) that tells somebody to come up with good reasons (Trivers, 1997; see also Trivers, 2000). Yet suppose there is a voice in someone's brain that whispers during a conflict: 'Are you sure you want to do that?' Since this voice is generated independently of the person's will, the decisions are no longer taken by the individual because a voice in the brain determines what someone does. Again, as in the case of the irresistible impulses, this is an example of psychopathology. For instance, one of the symptoms of schizophrenia is that individuals hear voices that order them to do something.

However, in the case of animals, are there examples of conflicts in the mind that can be explained in terms of brain mechanisms? Animals display conscious, goal-orientated behaviour and therefore there may be conflicts in concrete situations between different tendencies, or between tendencies and primitive emotions. These are manifested in the *hesitations* in their behaviour (see also Glock, 2000; 2003). If, for example, a subordinate chimpanzee sees a banana but also notices that he is being watched by the alpha male in the group, he hesitates to pick up the banana. Then we can state that there is a conflict between an inclination to pick up the banana and the fear of being attacked by the alpha male. We notice the hesitation in the chimpanzee's behaviour because he is moving around cautiously and since his gaze alternates between the alpha male and the banana. Although the chimpanzee does not have rational considerations, he can change his goals depending on what happens at that very moment (what the alpha male is doing, what the distance is to the banana, etc.). Does it make sense to explain the hesitations in the behaviour of the chimpanzee in terms of mechanisms in the brain without referring to the primitive considerations and options of the chimpanzee in this conflict situation?

There are examples of conflicts in animals where the influence of primitive consideration can be neglected. In these examples we do not observe hesitations accompanied by facial expressions (like surprise) and display of puzzlement (gestures like head scratching) followed by renewed activity. For example, ethologists have shown (in birds and fish) that conflicts between two opposite tendencies occur in the borderlands between two territories. If the owner of a territory is near to the centre of that territory, he will attack an intruder, but if the owner is at the border between two territories, the tendency to flee is activated too. These conflicts are visible in the behavioural movements of the animal (an animal cannot simultaneously attack and flee). Sometimes the animal alternates between movements towards and away from the opponent; sometimes he displays behaviour that appears to be a 'compromise' between attacking and fleeing; and sometimes a displacement movement is shown (a behaviour pattern irrelevant in the specific situation). In these cases behaviour is, as in the example of the turkey hen, almost explicable in terms of 'mechanistic movements'. According to ethologists, these conflicts between tendencies arise because two opposing circuits in the brain are simultaneous activated. Hence, there may be neurobiological mechanisms for the observable ambivalences in their behaviour (see, among others, Manning, 1972 [1967], chapter 5).

5.6   The crypto-Cartesian self and the Aristotelian agent

Evolutionary theorists assume that human behaviour can be resolved into two sources: genes and ideas. They have argued that our preferences and choices are influenced by genes and ideas since these are agents in the mind/brain. However, they also postulate, just as their (crypto-) Cartesian predecessors did, the existence of a Self. In situations where genes and ideas, as different internal voices, offer different advice, a Self needs to choose. Haig (2006, p. 21) has postulated that 'the Self can be viewed as the arbiter that mediates among the conflicting parties and then decides'. Humans are, according to Haig, free in the sense that genes and ideas do not determine our choices. In the wake of James he assumes that there are different agents in the mind/brain *and* supposes that there is a central Self mediating among agents. Yet what is the central Self that chooses in such situations, and how can we investigate its actions? Since this problem is left unexplained by evolutionary theorists, I will discuss James' ideas on what he called 'the Self of selves'.

Descartes thought that awareness of his thought proved that the thinking self is an (immaterial) substance. Although Descartes' conception of the self as an immaterial substance was rejected, philosophers like Locke and Hume and psychologists like James continued to speak of a self. James argued that within the stream of consciousness there is an active element in consciousness that he called the Self of all the other selves. This central Self is the source of effort and attention and the place from which the fiats of the will emanate. Although James, like the evolutionary theorists, did not think of the Self as an immaterial substance, his ideas on the Self may be characterized as crypto-Cartesian, since he continued to speak about the Self as a separate element (and this raises the problem of what this element is). James developed his ideas on how the Self resolves conflicts in the mind as an extension of his ideo-motor theory. According to this theory, images (or representations) are the prerequisite of voluntary movements. In simple cases an image, based on a perception, *causes* a volitional movement (James, 1950 [1890], vol. 2, p. 522): the act follows immediately after we have had an anticipatory image. For James insisted that a voluntary movement is in its very nature impulsive: the thought is directly correlated with an impulse on its way to instigate a movement. Only in cases in which we think of an act for an indefinite length of time without immediate action taking place may we talk about a will-force added to the image (James, 1950 [1890], vol. 2, p. 526). These are, according to James, instances of antagonistic thoughts (comparable to the voices of genes and ideas postulated by evolutionary theorists). Hence other ideas present in the mind may inhibit the impulsive power of a given motor image. These cases require what James called an inward effort to overcome indecision. Yet when the blocking is released as the result of deliberation, 'we feel as if an inward spring were let loose, and this is the additional impulse or *fiat* upon which the act effectively succeeds' (1950 [1890], vol. 2, p. 527).

However, James' crypto-Cartesian ideas suffer the same incoherence as their Cartesian predecessor. I will briefly discuss three problems. First, James thought, just as Descartes did, that voluntary actions proceed directly from the mind (see Table 5.2). For the mind is taken to be the agent that does something and brings about movements. Volitional movements are, according to James, initiated by a mental act performed within the mind. Yet how can we determine that a mental act initiates a movement? The problem is that these ideas cannot be empirically tested, since there are no criteria of identity of the mind. We do not know how to identify the mind as an active element in consciousness, how to measure this element, etc. And since there are no criteria of identity for minds, it is problematic to say that the mind has causal powers and can, therefore, interfere with physical processes. We would not know what kind of empirical evidence someone can provide if he or she argues that the mind causes a voluntary movement. For how can we determine that the mind causes this movement if we cannot identify this element? Second, James thought that intentional movements are *caused* by an impulsive power of a given image in the mind. He referred to impulses as acts of the mind because he wanted to develop a causal explanation of intentional movements. But an action caused by an impulse is not an example of an intentional action (see also above). If we want to do something, we do not have an impulse (or desire) in our mind that causes our movements. In the case that we want to do something, we have made a decision or developed a plan based on considerations. Hence we can give reasons for what we want. We have made a choice out of several options and may introduce our choice by saying 'I want' (e.g. to buy a book). By saying what we want we introduce a goal we are aiming at (Hacker, 1996a; see also Bennett and Hacker, 2011). And if someone says that she wants to do such-and-such, we cannot predict by what movements she will try to achieve her goal, just as we cannot predict the precise movements of a chess-player, although we can, perhaps, predict the move a chess-player will make (since we have, given the position of pieces on the chessboard, an idea of what his strategy is). Hence the idea that the mind initiates and causes movements and that the very nature of the will is an impulse is mistaken. Third, James' proof of the existence of the central Self makes no sense. He argued that its existence can be demonstrated by introspection since the Self is *felt*. Yet what is this feeling of the Self? James found through introspection that the Self of selves 'is found to consist mainly of the collection of . . . peculiar motions in the head or between the head and the throat' (James, 1950 [1890], vol. 1, p. 301). This absurd answer cannot be seen as adequate since no criteria for identifying these feelings are given. Since James misguidedly thought that the central Self is *in* us, he thought that introspection, conceived as a form of inner sense, will disclose its nature. Yet introspection is not an inner observation and it is, therefore, an illusion to think that introspection will disclose the central Self. There is no Self of selves in our mind/brain forming intentions based on deliberations. We, as rational agents, form intentions.

5.7   Conclusion: indecision instead of a divided Self

When evolutionary theorists investigate conflicts in the mind (or in the brain or mind/brain), they assume that there are different agents in the mind: instincts, ideas and a thinking Self. In this chapter I have criticized the conceptual framework used by evolutionary theorists since it provides an incoherent and misleading picture of the behaviour of humans and other animals. I have argued that the idea that there are mechanisms that cause 'antagonistic thoughts or voices' in the mind/brain is based on the crypto-Cartesian conception of the mind. The Aristotelian conception of the mind, I suggest, provides a clearer picture of conflicts in the mind: these conflicts are resolved by humans acting as rational agents on the basis of considerations and deliberations, and not by mechanisms or agents in their brains or minds.

When instinctive and goal-directed behaviour gradually transforms into reasoned behaviour as children learn a language, they become individuals or persons who are answerable to reasons and responsible for their behaviour. Their initial instinctual responses become, with maturation, under the control of reason and reflection. It is, of course, possible that an individual has conflicts, but it is ultimately the individual who reports these conflicts and tries to solve them, not her brain or a Self. The mature individual is not divided in the sense that there are entities in her mind that cause actions independently of the individual. Although someone may feel passions or may have an uneasy conscience (because she has internalized ideas), the conflicts that may arise are her conflicts and it is the individual who tries to solve them. If someone describes passions, tendencies and impulses as 'entities' that cause her to act in a certain way and on which she has no grip, then there is something wrong. Passions or impulses are, in the case of the human species, not separate parts of the individual because our natural behaviour undergoes transformations when children learn a language. The reason why evolutionary theorists invoke the idea of the divided Self has a special background. They have developed their ideas in opposition to social scientists who defended a one-sided, rational view of humans. Rightly they have insisted on the fact that humans are more than rational beings: humans take care of their children, search for food, have an emotional life, etc. But this observation does not imply that human rationality is an entity in our mind or brain that is separable from our instinctive life.

Within a coherent, Aristotelian framework our rational life is seen as an extension of our instinctive life. Like animals, humans may be hungry, but unlike animals, for children who have mastered the rules of a language, the fact that dinner is being served within half an hour may make the hunger less uncomfortable (although, of course, they still have to eat something to stop the sensation). Such a thought does not occur in the animal mind. This is not because a thinking substance is only present in the human species, or because humans have more instincts than animals. It is because the child has learnt to express his or her bodily sensations verbally and can, therefore, act in conscious response to a bodily desire, that is, hunger. And if he or she acts out of a desire, the child learns to refrain temporarily from satisfying the desire.

The conceptual framework used by evolutionary theorists is also a source of misleading ideas about animal behaviour. Animals may have primitive consideration when they face conflicting situations because they develop complex, goal-directed behaviour. This is the case if animals are able to learn by insight (thus through a more complex learning process than conditioning). Insight learning may be the background of forms of primitive knowledge in the case of animals. Although animals do not display rational, intentional behaviour, this does not mean that they only have instincts and learn according to the rules of classical and instrumental conditioning. We can notice their primitive considerations by observing their hesitations, their facial expressions and changes in their behaviour when they appear to have solved a puzzle. It comes as no surprise that these primitive forms of knowledge are the background of different cultures in populations of chimpanzees. Since human, rational behaviour evolved as an extension of abilities already present in the forerunners of humans – bonobos and chimpanzees – we can expect primitive forms of intentional behaviour in these apes (although these animals have taken a different route and, hence, the risks of anthropomorphizing their behaviour are clearly present). Humans display intentional behaviour, but the development of intentional behaviour was only possible if the forerunner of humans displayed complex, goal-directed behaviour.

If the analyses given in this chapter are correct, then the conclusion must be that a framework combining an Aristotelian view and evolutionary theory is better for explaining the behaviour of animals and humans. This framework explains why only humans evolved into rational creatures, namely that they developed a language that allowed them to give reasons for their actions. It also explains why primitive forms of intentional behaviour may be present in apes, since they display behaviour also displayed by the forerunner of the human species. And it makes clear that there are no conflicting agents in the mind causing intrapsychic conflicts, for so-called 'intrapsychic conflicts', save in cases of psychopathology, are not caused by agents in the mind but reflect indecision.

CHAPTER 6

Weismann, Wittgenstein and the homunculus fallacy

6.1 Introduction

Instincts fascinate biologists, psychologists and philosophers for two interrelated reasons. First, instincts seem to guide organisms through a complex natural and social world, and contribute therefore to the harmony between an organism and its environment. Yet how can we understand this apparent harmony between instinct and environment? Two opposing explanations are distinguished (see Jablonka and Lamb, 2005). Neo-Darwinists suggest that instincts evolve as the result of natural selection and reject Lamarck's theory on the inheritance of acquired characteristics. The harmony between organisms and their environment is, according to them, explicable as an evolutionary adaptation. Since instinctive behaviours evolve as the result of selection on genetic variation, they are, according to neo-Darwinists, innate rather than learnt. They develop during the lifetime of an individual as the result of decoding of information in the genome. Neo-Darwinists assume that future investigations will provide more insight into the nature of this 'phylogenetically acquired information'. Neo-Lamarckians, however, argue that instincts may *also* evolve as the result of instruction. They argue that learning processes are involved in both the development and evolution of instincts. Neo-Lamarckians assume that future studies will provide insight into how instinctive behaviour displayed by organisms in the present generation is, in part, the product of the accumulated experiences of past generations fixated in developmental programmes. Second, instincts appear to involve knowledge, for animals seem to know what an appropriate act is in a given situation. For instance, newborn animals instinctively appear to know how to (re)act in a certain situation, since they immediately use their limbs in a purposeful manner and act without hesitation. As Bonner (1974, p. 47) put it, 'the animal is born with a memory of how to do something that it has never done before'. Again, it is assumed that future studies will show how animals are able to act on the basis of this knowledge stored in the DNA or in developmental programmes. In the case of the human species, these investigations are extended with studies into the language instinct (see Pinker, 1994). Hence the problem arises how we can investigate the apparent harmony between linguistic behaviour and the world as an extension of the harmony between instinct and the world.

In this chapter I shall discuss the contributions to our understanding of the harmony between instinct, knowledge and the world of the evolutionary biologist August Weismann and the philosopher Ludwig Wittgenstein. The reason why I discuss their contributions is that both have discussed homunculus fallacies. Weismann criticized Lamarck's theory of evolution since Lamarck presupposes, according to Weismann, the existence of a homunculus in germ cells; Wittgenstein criticized theories that explain the apparent harmony between knowledge (or language) and the world by referring to information stored in the mind, brain or DNA, since these theories mistakenly assume that cognition can be attributed to parts of a living organism (as if these parts are homunculi). I argue that Wittgenstein's ideas (which are elaborations of an insight discussed by Aristotle; see Chapter 5, section 5.2) can be seen as an extension of those of Weismann. Hence the combination of their ideas, I suggest, sheds new light on the old problem of the harmony between instinct, knowledge and the world.

6.2   Weismann on instinct

Weismann is well known because of his critique of the idea of Lamarck (accepted by Darwin) that instincts arise and decline from habit in use and disuse. In his critique, Weismann focused on the problem of how an acquired habit can become heritable. An acquired habit affects the somatic cells of the body, but in order to become heritable, there have to be changes in the germ cells as well, since offspring originate from the germ cells. It is, according to Weismann, possible that the environment and the somatic cells of an organ affect the determinants in germ cells (residing in the chromatin of the cell nucleus and necessary to produce an individual), but he doubted that these influences induce a change in the determinants corresponding to the change in the somatic cells. Weismann formulated the theoretical problem as follows: how can new acquirements of the personal part have representative effects on the germinal part?

Weismann discussed the effects of memory-exercises of an actor as an illustration:

How could it happen that the constant exercise of memory throughout a lifetime, as, for instance, in the case of an actor, could influence the germ-cells in such a way that in the offspring the same brain-cells which preside over memory will likewise be more highly developed – that is, capable of greater functional activity? (1983 [1904], vol. 2, pp. 107–108)

Proponents of Lamarck's theory assumed that the circulatory or nervous system acted as the intermediary between somatic cells and germ cells. Hence the problem arises of how the states of the brain cells after memory-exercises are 'communicated by the telegraphic path of the nerve-cells to the germ-cells' and, subsequently, how in the germ cells only the determinants of the brain cells are modified in such a way that, in the subsequent development of the offspring, 'the corresponding brain-cells should turn out to be capable of increased functional activity'. The problem noted by Weismann is that the determinants in the germ cells are assuredly different from brain cells:

> But as the determinants are not miniature brain-cells ... but only living germ-units which, in co-operation with the rest exercise a decisive influence on the memory-cells of the brain, I can only compare the assumption of the transmission of the results of memory-exercise to the telegraphing of a poem, which is handed in in German, but at the place of arrival appears on the paper translated into Chinese. (1983 [1904], vol. 2, p. 108)

Hence Weismann, I conclude, criticized Lamarck since his theory mistakenly presupposes a homunculus in the germ cells. I summarize this as the *empirical homunculus fallacy*. This needs some clarification.

First, I use the term 'empirical homunculus fallacy' to signify Weismann's description of a mistaken empirical assumption in Lamarck's theory. The fallacy is not that Lamarck committed a logical error but a biological one. Weismann criticized a *ceteris paribus* assumption in Lamarck's theory: the theory on the inheritance of acquired characters tacitly presupposes the existence of miniature memory cells (i.e. a homunculus). If one could demonstrate the existence of these miniature memory cells, then these cells could perhaps be transformed in a similar way to the memory cells in the brain of the actor. Yet the determinants in the germ-plasm are, according to Weismann, not miniature brain cells.

Second, Weismann did not discuss a conceptual problem present in the theory of Lamarck, namely how we can understand the role of memory cells in the retention of knowledge. Does this mean that knowledge is stored as the result of memory exercises in cells or structures in the brain? Answers to this question require, according to Wittgenstein, conceptual investigations, and in this context he has discussed – what may be called – the *conceptual homunculus fallacy*. Hence the empirical error pointed out by Weismann should not be confused with the conceptual error discussed by Wittgenstein (namely that it is senseless to say that knowledge is stored in brain cells or germ cells), although both fallacies apply to the example of the actor discussed by Weismann. Wittgenstein's conceptual critique is, therefore, an extension of Weismann's empirical objections.

The upshot of Weismann's argument is that instincts did not evolve as the result of habit in use and disuse, but evolved as the result of natural selection. Weismann underpinned this (neo-Darwinian) insight with many empirical observations. Against those who believed in the inheritance of use and disuse, he presented examples of instinctive behaviours that cannot have been learnt since they were immediately displayed by organisms as soon as they were born. Furthermore, he pointed out that some instinctive behaviour is geared to specific stages of the life cycle and performed only once. Therefore it cannot be the result of training or practising. As to the problem of what instincts are in terms of neurophysiology and how they evolved during the course of evolution, Weismann argued that they evolved out of reflexes. Initially they were 'complex reflexes' and later evolved into unconscious, goal-directed activities. His main argument for thinking that instinctive behaviour evolved out of reflexes was that just like a reflex, instinctive behaviour may be liberated by a single stimulus. Yet in contrast to reflexes, instinctual behaviours consist of concatenated series of movements. They are, according to Weismann, examples of goal-directed action, since there is a sequence discernable in the movements. Instinctual behaviours end with what ethologists later called a consummatory act. Weismann discussed many examples to illustrate the idea that instinctive behaviour is unconscious, goal-directed behaviour.

> All around we can see that animals know how to use their parts or organs in a purposeful manner: the duckling swims at once upon the water; the chicken which has just been hatched from the egg pecks at the seeds lying on the ground; ... and the predatory wasp requires no instruction to recognize her victim, ... she knows how to attack it, to paralyse it by stings, and then hesitates not a moment as to what she has to do next. (1983 [1904], vol. 1, p. 141)

According to Weismann, these instincts guide animals towards ends without having any consciousness of these ends. However, when we say that animals 'know' how to use the limbs of their body, we do not mean that instincts inform them how to do this. Weismann (1983

[1904], vol. 1, p. 146) quoted Lloyd Morgan who wrote in *Habit and instinct* about the pecking behaviour of chicks: 'It does not pick at the seeds because instinct says to it that this is something to be picked up and tested, but because it cannot do anything else.' Hence instinctive behaviour is, according to Weismann, displayed without reasoning. If instincts were inherited habits (and amenable to reason), as Lamarck and others had argued, then there should be evidence of the possession of 'that degree of intelligence which would have induced the variation in the previous habit, that is, in manner of movement' (Weismann, 1983 [1904], vol. 1, p. 155). However, since instinctive behaviours are displayed stereotypically and without hesitation, this showed to Weismann that they are at first performed without reasoning. In 'higher' animals instinctive behaviour can, according to him, be modified or inhibited by intelligence. In humans instinctual behaviour transforms into intentional behaviour. Weismann discussed breastfeeding as an example. The suckling of a baby is at first an instinct: the child seeks about with its mouth. Yet if the sucking 'is continued into the second year of life, as not infrequently happens in the southern countries of Europe', it is an example of primitive intentional behaviour, since the 'child knows exactly why it wants the breast' (1983 [1904], vol. 1, p. 143). The child can express its bodily sensations verbally and can ask for the breast if it is hungry. It no longer seeks the breast instinctively but acts in conscious response to a bodily desire, i.e. hunger or thirst. And if it acts out of a desire, the child learns to refrain temporarily from satisfying the desire.

Just as it is senseless to say of hatchlings that instinct informs them how to peck, it makes no sense to say of a baby when it seeks about with its mouth that it 'knows' what it does or that it possesses the concept 'milk'. Just as the chick, the baby acts without hesitation and does not doubt whether the fluid produced by the breast is, actually, milk. For doubt, ignorance, etc. are, during the early developmental stages, logically excluded (as I will argue below). In humans, instinctive behaviour is, of course, rapidly transformed into intentional behaviour when children start using a language. Linguistic behaviour is an extension of instinctive behaviour and enables children to develop intentional behaviour. Linguistic expressions like 'Want milk', 'I am hungry' or 'I am thirsty' are extensions (and, in part, replacements) of instinctive behaviours. Just as with the instinctive precursors, these expressions are at first still done unhesitatingly and without reasoning. These expressions are new forms of behaviour but are no more due to reasoning then the preverbal, instinctive ones. Hence these linguistic utterances can also be called immediate expressions to emphasize the lack of reasoning and thinking. For that reason Malcolm has called these expressions instincts in the secondary sense (Malcolm, 1982). For the behaviour of a child is, like instinctive behaviour, at first not based on beliefs and thoughts. It is acting without any propositional thought.

6.3   Innate or learnt

Within the confines of the conceptual framework discussed by Weismann, there are still some conceptual (and, of course, many empirical) problems to solve. For instance, the adaptiveness of instinctive behaviour was used by Weismann as an effective argument against the theory of Lamarck. Yet if instinctive behaviours are selected adaptations and if the determinants underlying these behaviours are transmitted from one generation to the next through the immortal germ line, does this imply that there is a conceptual distinction between innate, instinctive behaviour and acquired, learnt behaviour? The ethologist Lorenz argued that there is indeed a sharp distinction between instinctive and learnt behaviour. Lorenz developed his ideas in opposition to behaviourists who thought that animal behaviour is largely learnt. He distinguished two types of behaviour: instinctive behaviour, i.e. species-specific behaviour that has been 'acquired' during the evolutionary history of the species (phylogenesis); and learnt behaviour, acquired during the development of the individual (ontogenesis). Yet his ideas were criticized by both ethologists and psychologists. According to the critics the argument from adaptiveness does not prove that instinctive behaviour is fully innate. For example, the psychologist Lehrman argued that, although instinctive behaviour may have been selected during the course of evolution, it does not follow that learning processes are excluded during the ontogenesis of instinctive behaviour. Since Lorenz, just as Weismann, had emphasized that instinctive behaviour is immediately displayed by an individual after it is born, Lehrman (1953) argued that even in these cases it is still possible that organisms may learn certain (sensory or motor) components of instinctive behaviours during prenatal stages. For instance, based on research of the behaviorist Kuo, Lehrman suggested that portions of the pecking behaviour displayed by a chick after hatching could have been learnt in the egg. Since the head is rhythmically moved up and down through the beating of the heart, this self-induced 'experience' may contribute to the development of the pecking behaviour. In response to this critique, Lorenz stated that the argument of Lehrman is unsound, for:

It . . . remains unexplained why only certain birds peck after hatching, while others gape like passerines, dabble like ducks, or shove their bills into the corner of the mouth of the parents as pigeons do, although they all, when embryos, had their heads moved up and down by the heartbeat in exactly the same fashion. (1965, pp. 23–24)

Hence Lorenz was hardly impressed by the critique and still believed that the conceptual distinction he made between innate and acquired behaviours is warranted. Again, he used phylogenetic arguments to substantiate his ideas. But does his response warrant a conceptual distinction between innate and acquired behaviour? According to the critics the problem remains that Lorenz apparently confuses phylogeny and ontogeny. Evolutionary arguments alone cannot, according to them, decide whether instinctive behaviours are innate. Whether behaviour is innate requires *also* an investigation into the development of behaviour. Although the critics (see Lehrman, 1970) admit that in the specific case of the pecking behaviour the empirical arguments for saying that self-induced 'experience' plays a role are weak (see, for example, Hamburger, Wenger and Oppenheim, 1966), this does not prove that every instinctive action or response is fully innate. Furthermore, why should we presuppose the exclusive disjunction in our argumentation, i.e. that behaviours are either innate or learnt? For logic does not exclude the possibility that during ontogenesis experience may mould developmental processes.

Investigations of the psychologist Gottlieb (1972, 1992) later demonstrated that prenatal experience can play a role in the development of behaviour. He has shown that the vocalizations made by embryos of the Peking duck (*Anas platyrhynchos*) in the egg contribute to the development of their ability to discriminate the maternal call. Hence their instinctive response to the maternal call is, in part, moulded by self-stimulation. If the syringeal ('voicebox') membranes are manipulated in such a way that the embryo is prevented from vocalizing during prenatal stages, the discriminatory abilities, although still present, do not develop as well as in controls. It appears that the duckling's ability to respond selectively to the maternal call of their species develops in the complete absence of prior exposure to normally auditory stimulation of a vocal nature, but it does not become or does not remain fully developed under such circumstances. An evolutionary explanation for the role of self-stimulation in this developmental trajectory of the ducklings is that it is important for the hatchling to discern the call of their mother directly after hatching, since the poikilothermic ducklings run the risk of dying because of cold (and predators). So the developmental trajectory is adaptive, although the example does not fit the ideas of Lorenz on instinctive responses, since experience (self-stimulation) contributes to a putative innate response.

The conclusion to be drawn from the discussions between Lorenz and his critics is that it is misleading to use a conceptual dichotomy between innate, instinctive and learnt behaviour. Instinctive and learnt behaviours are better conceptualized and studied as the two poles or extremes of a continuum. In between are examples showing that it makes less sense to separate genetic and learning components. Hence it is for methodological reasons better to avoid the nature/nurture dichotomy. For this dichotomy misleadingly suggests that experience is unimportant during the development of instinctive behaviour and underestimates the possible adaptive role of, for example, self-stimulation. It is, as a research strategy, better to assume a continuum in order to get a full measure of the possibilities.

Whether learning is involved in the development of instinctive behaviour will depend on the social and ecological environment organisms are living in. For obvious reasons the cuckoo and cowbird have been used as examples to argue that the ability to recognize co-members is in these species probably fully innate. These are parasitic birds that lay their eggs in the nests of other birds (acting as host parents). After they mature, females respond to the song of a male by adopting a 'copulatory posture'. Studies have shown that cowbirds, reared in isolation, display the same response (King and West, 1977). It is therefore assumed that the response is innate, in contrast to the host species of cowbirds in which learning is involved. The selective advantage of a genetically determined pattern is obvious in this case. Nevertheless the problems mentioned by the critics of Lorenz remain: a deprivation experiment is not sufficient for showing that the response is fully innate, for one still has to demonstrate that experiences during prenatal stages do not affect the development of the response. Yet if the response of the cowbird is fully innate, then this will be reflected in the regulation of the mating behaviour of the females. One can predict in that case that the auditory parts of the brain, when these receive a particular sound signal, will automatically send impulses to those parts of the brain that regulate reproductive behaviour. The connections between the auditory and motor parts will probably be established by genes during the development of the brain. By contrast, it is likely that in the host species, in which imitation of songs is a prerequisite for the development of reproductive behaviour, other parts of the brain and processes like synaptic plasticity are involved in the development of mating behaviour.

There are, of course, other experimental possibilities with which to disentangle the possible role of brain cells and structures in the development of instinctive behaviours. I will briefly discuss one example. Balaban (1997) implanted parts of a quail midbrain into chick embryos, creating animals known as chimeras. The chimeras moved their heads like chickens but crowed like quails. This shows that not all transplanted cells started to participate in the neural circuits of the (host) chick brain. The experiment also demonstrates that two components of the crowing behaviour, head movement and sound, have their origin in separate neural circuits. It is likely that these components have a different evolutionary history and did not evolve together as a unit. Other experiments have shown that the timing of effects has a major influence on the results: the later in the development the cells are transplanted, the more these cells reflect the source rather than their destination.

6.4 Instinct and the central dogma

After the discovery of DNA replication and the processes of transcription and translation involved in protein synthesis, Weismann's ideas have been reinterpreted in terms of molecular genetics (see, for instance, Maynard Smith, 1986, chapter 2). There is a specific reason why Weismann's arguments were reconsidered. Weismann thought that when embryonic cells divide each daughter cell receives a different set of determinants, and used this hypothesis to explain cell differentiation (the development of 'personal parts'). But molecular genetics demonstrated that the same genes reside in each cell. Hence the problem arises whether alterations in genes, residing both in memory cells and in germ cells, can explain the inheritance of acquired characteristics. According to Maynard Smith, Crick's central dogma (captured in the principle: 'information flows from DNA to proteins but not from proteins to DNA') shows why Weismann was right. Functional changes in memory cells primarily alter the role of proteins in brain cells (and, in some forms of learning, affect gene expression in brain cells without altering the sequence of the nucleotides). Hence the discovery of the central dogma is seen as a confirmation of Weismann's and as a refutation of Lamarck's ideas.

Maynard Smith emphasizes that Weismann's distinction between the germ and soma line not only reflects a division of labour between cells contributing to and foregoing reproduction, but also a division of labour between the roles of nucleotides and proteins in a life cycle. Nucleotides are selected to have properties ensuring accurate replication: the fit between complementary bases of the nucleotides prevents advantageous genotypes from getting lost. Proteins are selected for their role in the survival of the organism. Hence they have catalytic properties needed for growth, metabolism, the regulation of behaviour, etc.

Although it looks as if the rise of molecular genetics marked the permanent end of Lamarckian explanations of instincts, there are many investigators who think otherwise. Based on knowledge developed in molecular genetics and later epigenetics, they have rejected Weismann's ideas and defended neo-Lamarckian ideas as a sound alternative (see, for example, Jablonka and Lamb, 2005). How can we understand that the same data (generated by genetic studies) result in two opposing views? At face value, the central dogma seems to provide a strong argument for the innateness of instincts. A simple reading of the central dogma says that evolution and development, phylogeny and ontogeny, are connected since the genetic code is involved in both processes. Genetic information is replicated during cell division and transmitted through the germ line from one generation to the next, and the same information is transcribed and translated during the development of an individual. This explains why instinctive behaviours are innate. But, as I have discussed above, there is also evidence that experience contributes to the development of instinctive behaviour. Hence there seem to be two sources of information that are both involved in the development of an organism. And since evolution and development are connected, one can ask whether evolutionary processes can also include Lamarckian instruction processes. Neo-Lamarckians believe that this is indeed possible. They argue that, just as innate and learnt behaviours are better seen as the two poles of a continuum, neo-Darwinian and neo-Lamarckian evolution are also the extremes. One extreme is that instinctive behaviours evolve as the result of purely (Darwinian) selection on random genetic variation; the other is that instinctive behaviours evolve from a purely (Lamarckian) instruction process. 'Between these extremes we find the majority of the processes in the real world, which are to varying degrees both instructive and selective' (Jablonka and Lamb, 2005, p. 102).

The connection between development and evolution can also be explained in another way (see Bonner, 1974, p. 5). Development is a stage in a life cycle. Life cycles are connected one to another by a single-celled stage. But if one looks at a whole series of life cycles which change through the course of time, then one has evolutionary development. Hence it looks as if the causal chain of adaptive evolution begins with development (but, as Bonner immediately added, natural selection can only operate if there is variation and reproduction). Changes during development are then the source for evolutionary changes. This picture of the connection between development and evolution is the reason why neo-Lamarckians assume that as the result of instruction (during development) information is somehow and somewhere stored in the organism so that it is available for future generations. Haig summarized the picture as follows: there is a two-way flow of information between genotype and phenotype. 'Genotypes interact with the environment to produce phenotypes and then phenotypes interact with the environment to determine which genotypes are replicated to become the focus for the next round' (Haig, 2007, p. 426). Because of this two-way flow of information, it is, according to neo-Lamarckians, possible that instinctive responses originate in instruction processes if these instructions are transmitted to the next generation. They have used the following, hypothetical example (originally discussed by Spalding 1954[1873]) as an illustration of their ideas. Suppose that Robinson Crusoe notices that the parrots on the island are very good imitators. He conceives the idea of teaching them some English words. After a certain period he successfully teaches parrots to say: 'How do you do, sir?' Now suppose that, after Robinson Crusoe died, the parrot parents take over his role as teacher so that the behaviour is then transmitted from parents to offspring. If there is genetic variation affecting this learning process (the best learners hardly need an instruction), then it is possible that there is Darwinian selection. For if the heritable variation correlates with reproductive success (the learning process is costly in terms of fitness), there will be selection in favour of fast learners. The result may be that, eventually, so little learning is needed that someone who is visiting the island and observing the parrots thinks that their behaviour is an example of instinctive behaviour.

There is also a neo-Darwinian version of the story. According to this version, Robinson Crusoe uses a breeding programme. Suppose that there is genetic variation and that some parrots spontaneously say 'Sir'. If Robinson Crusoe only uses these birds for breeding and if, because of a second mutation, birds evolve that also say 'do', then it is possible that some birds say 'Do sir'. The process may go on and in the end Robinson Crusoe has bred parrots saying 'How do you do, sir?' When Robinson Crusoe dies and we visited the island, we would be surprised to find this instinctive behaviour.

Both versions are possible explanations of the parrot's behaviour. But do the neo-Lamarckian version and neo-Darwinian version both result in instinctive behaviour? Jablonka and Lamb think so, but I believe that their argumentation is mistaken. My main argument is that an answer to this question does not depend on the role of 'learning', but on the presence or absence of *cognition* in putative instinctive behaviour. I will use again the example of parrots 'learning' to say 'How do you do, sir?' to illustrate my argument.

First, I assume that the breeding experiments result in innate behaviour, for there has been selection on genetic differences causing phenotypic differences. Of course, experience may be involved in the development of the behaviour, just as self-stimulation is involved in the development of the discrimination response of the ducklings. But we call the behaviour of the parrots instinctive in this case if they do not *understand* what they say, i.e. if cognition is excluded. In Weismann's terms, we will say that the parrots, after the breeding programme, display the behaviour unhesitatingly, spontaneously and without propositional thought. Second, the outcome of the instruction-experiment depends on whether cognition is involved. If the parrots understand what they say, then they can use this knowledge and can be said to have acquired knowledge. In that case they know the difference between saying it correctly and incorrectly (perhaps they know the difference between saying 'How do you do, sir?' and 'How you do do, sir?'), although it will be hard for us to find out whether they use knowledge since the birds cannot tell us what they know. But if knowledge is involved, then their behaviour is not an example of an instinct according to Weismann's criteria. For (1) the saying is not displayed unhesitatingly, spontaneously, etc. and (2) a (brief) instruction period is still required for the acquirement of the knowledge. The other possibility is that the parrots do not understand what they are saying: they have learnt the saying through, for example, conditioning. In that case we can imagine that, through selection on quick learners, genes can 'take over' a large part of the instruction process. Hence the behaviour is in the end largely spontaneous (and hardly requires instruction). This case is, then, to a certain extent comparable to the role of self-stimulation in the development of the discrimination response of ducklings, although self-stimulation in the case of the ducklings accelerates an originally innate response, while genes in the case of the parrots accelerate a learning process. It may be said that the parrots have an instinctive disposition to learn the saying.

This interpretation of the parrot story does not violate Weismann's ideas. The reason is that the crucial distinction made by Weismann is the distinction between instinct and cognition. Weismann criticized Lamarckian explanations of instinctive behaviours and made a conceptual distinction between instinctive behaviour, explicable in terms of evolutionary theory, and behaviour based on acquired knowledge. In order to characterize instinctive behaviours, he used concepts such as 'hesitation', 'stereotypically', 'species-specific', 'goal-directed', 'innate', etc., and contrasted these concepts with the concepts we use to describe behaviours based on knowledge, such as 'intentional', 'learnt', based on 'memory or reasoning', etc. There are, of course, learning processes (such as self-stimulation or conditioning) which contribute to the (accelerated) development of behaviour. But these do not involve cognition. Hence one can easily accommodate these learning processes to neo-Darwinian theories, as the discussion of Spalding's hypothetical example demonstrates.

Fundamental for future discussions of Weismann's conceptual framework is therefore the distinction between instinctive behaviour and acquired behaviour based on knowledge. I will argue that a better understanding of this distinction requires an extension of Weismann's critique of the empirical homunculus fallacy with Wittgenstein's discussion of the conceptual homunculus fallacy.

6.5   Facts of mind and matter

The Lamarckian Spalding (1954[1873], p. 8) believed that 'instinct in the present generation is the product of the accumulated experiences of past generations'. However, Spalding noted that this hypothesis is not generally appreciated for conceptual reasons. The reason why investigators doubt the validity of the hypothesis is, according to Spalding, that the mind is involved:

> The facts of mind ... differ from material things in this important respect, that whereas the latter can be stored up, volitions, thoughts, and feelings, as such, cannot. Facts of consciousness cannot be thought of as packed away like books in a library. (1954 [1873], p. 8)

Hence the problem arises how the facts of mind become permanent if one cannot store thoughts or memories in the mind.

Spalding developed the same solution to this problem as his empiricist predecessors. He assumed that the facts of mind depend on the 'corresponding impress given to the nervous system'. There is, according to Spalding, for every fact of mind a corresponding fact of matter. Hence the facts of mind are, given the material fact, the same in parent and offspring, for the material fact, whether produced by repeated experiences in the life history of the individual or inherited from parents, guarantees that the corresponding mental fact will be the same. It is interesting to note that Spalding (1954 [1873], pp. 8–9) tried to convince his readers by putting the following question to them. Suppose that we form by some mysterious process an organism, the same in every particle and fibre as your friend, such that outwardly you cannot tell the one from the other. Will the newly created man, by virtue of his identical material organization, not possess the personal identity of the other? I will return to this question below.

Spalding did not discuss the problem how changes in the states of the brain cells are, as Weismann put it, 'communicated by the telegraphic path of the nerve-cells to the germ-cells'. But other Lamarckians later discussed precisely this problem (see Weismann, 1983 [1904], vol. 2, pp. 109–112). For example, Hering assumed that a germ cell experiences in itself all that befalls an organ (as the brain or a limb) and argued that these experiences stamp themselves upon substances in the germ cells, just as sense-impressions or perceptions stamp themselves upon the nerve-substance of the brain (what Spalding called the material fact). And just as the brain is able to bring memory-pictures (what James later called traces and what nowadays are called representations) back to consciousness, Hering thought that the experiences acquired by germ cells are later reproduced during the development of the germ cells (i.e. in the offspring).

In his comments on the ideas of Hering, Weismann discussed only the problem of whether there are empirical arguments in favour of this theory. He concluded, correctly, that there is no empirical evidence. Hence the inheritance of functional modifications is not a fact but, according to Weismann, an interesting but far-fetched analogy with the retention of information in memory. But there is another (and, as I will argue, mistaken) assumption in Hering's argumentation: just as empiricist phil-osophers assumed that impressions are *stored* in the brain/mind (and are therefore *available* to someone), Hering believed that the germ cells underwent a process comparable to receiving and storing knowledge in the brain: the (chemical) substances in the germ cells contain the stored information as the result of the modifications of the substance in the germ cells. This mechanism postulated by Hering raises the conceptual problem of whether it makes sense to say that knowledge is stored in cells as the result of an impression stamping itself on a substance. An answer to this question requires, first, a longer conceptual analysis of the differences between instincts, learning through self-stimulation or conditioning and acting based on acquired knowledge, and second, an analysis of what we mean by the retention of knowledge.

6.6   Instinct, knowledge and abilities

We make a conceptual distinction between innate and acquired abilities (see Bennett and Hacker, 2003, chapter 5). Examples of innate abilities are the ability to breathe, to perceive and to move one's limbs. The ability to walk is an acquired ability, for we learn to walk. Yet although it is an acquired ability (and an example of a two-way ability, for we can choose whether we will walk), learning to walk does not involve the acquisition of knowledge. We do not possess knowledge after we have learnt to walk. Learning to walk results in being able to walk, but there is no such thing as knowing or remembering how to walk, and we cannot explain what knowledge is acquired when we learn to walk. Hence when we lose the ability to walk as the result of paralysis, we do not forget how to walk, for there is no such thing as forgetting (and remembering) how to walk. As a corollary, another person cannot remind us how to walk.

There are, of course, many examples of acquired abilities involving knowledge (take Weismann's example of an actor). Knowing how to do something is then the upshot of learning or teaching. It results in knowing that, how, when, whether, etc. If someone knows how to do something, then it makes sense to ask of such an act that one remembers or recollects how to do it. And if someone who has learnt how to do something is not able to do so, then we will ask whether he has forgotten how to do it. It is possible that we can remind him how to do it (e.g. a prompter who aids an actor who has forgotten his lines or gestures).

Knowledge in these cases may be acquired through perception or through reasoning. It may also be acquired from the written (reading a book) or spoken word (attending a lecture). When we have acquired the knowledge of how to act in a given circumstance, then we can do something that shows that we are able to act. For that reason knowledge and the retention of knowledge are said to be *ability-like* (see White, 1982). To possess knowledge is, therefore, not to be in a *mental state*. By contrast, when I am angry or anxious then I am in a mental state. For anger and anxiety are emotions that have duration, that can be interrupted (by, for instance, sleep) and that may be more or less intense. We typically ask someone *why* he is in a mental state, for example, 'Why are you angry or anxious?' Possession of knowledge, however, cannot be interrupted (one does not cease to know while one falls asleep), knowledge does not have duration (it makes no sense to say that I knew how to integrate or differentiate last week, although I can, of course, lose my mathematical skills) and knowledge does not have an intensity. We find out that someone possesses knowledge not by asking why but by asking *how* someone knows. Whether he knows something is determined by observing *what he can do* (Bennett and Hacker, 2003, chapter 5). Knowing how things are or how something is done is being able to explain something or show how it is done. And whether someone remembers something is

determined by observing his ability to retain it, i.e. whether he can retell
and explain something or can still exercise an acquired ability. To forget
something is not to cease to be in a certain state, but to cease to be able to
do something.

Acts based on acquired abilities involving knowledge are, as Weismann
(1983[1904], vol. 1, p. 155) observed, characterized by the 'manner of
movement'. What is meant by this phrase? If one knows how to do
something, then it makes sense to say that one knows what it is to make
a mistake. For to know how to do something is to know the way to do it,
and to know the way to do it implies knowing the difference between
doing it correctly or incorrectly. An animal cannot tell how to do something (in)correctly, but if it acts out of knowledge, then its behaviour will,
in contrast to instinctive behaviour, be flexible. The flexibility is exhibited
in the plasticity of its response to circumstances, its recognition of error
when it occurs and its rectification of error in performance (Bennett and
Hacker, 2003, pp. 150–151). Hence acts based on knowledge are in this
sense characterized by the 'manner of movement'.

If I know that things are such-and-such, then I can act on the basis of
that information. This information provides me with reasons for doing
something or for pursuing a goal or project. Hence this information
enables me to form intentions ('I want to go to the cinema tonight because
I want to see Al Pacino in …'). Other animals know things and may
develop plans too (a lion may develop the plan of chasing a zebra after she
has located one), but since they are not language-using creatures, they
cannot give reasons for their plans and, hence, develop intentions only in
the attenuated sense (see further, Rundle, 1997, chapters 3 and 4). When
we want to determine what an animal knows, we have to investigate what
it can do. And what it can do is what it exhibits in non-verbal behaviour.
For example, a dog can be said to know where its home is and this
knowledge is exhibited in its ability to find its way (without getting lost);
a jay can be said to know where it has cached food if it is able to go to the
right place without hesitation. Hence animals also possess a mind, i.e. they
have, just as humans, the capacity to acquire intellectual abilities and to use
these acquired abilities (see Kenny 1975, chapter 1). And they have a will
since they develop goal-directed activities and have preferences. But the
main difference between animals and humans is that humans have an
innate capacity to acquire a language. Only humans are therefore responsive to reason.

Since humans are language-using creatures, the human mind also
includes the capacity to use symbols. For this reason humans can pursue

self-selected goals with the help of symbols and this enables them to develop purposes beyond the temporal and spatial present ('I want to go on holiday to Spain in May because of the nice weather'). Hence the combination of the intellect and will enables humans to pursue self-selected goals that go beyond the immediate environment in space and time.

Both instinctual behaviour and acts in which retained information is used are described by saying that an organism displays behaviour 'without hesitation'. Does this license the conclusion that in both examples knowledge is involved, i.e. innate knowledge in the case of instincts and acquired knowledge in the case of retained information? Knowledge, as was already noted by Weismann, is in instinctual behaviour excluded for logical reasons. When an animal displays instinctual behaviour unhesitatingly, then it, as Lloyd Morgan put it, 'cannot do anything else'. It does not hesitate or doubt whether the performed action is appropriate in the given situation, for doubt and ignorance are logically excluded. By contrast, if a dog goes to the right place where it has buried a bone without hesitation (it does not search for a long period), then we can say that it is certain about where it has hidden the bone (the dog does not hesitate because it knows for sure where it buried the bone). There are several empirical and conceptual arguments that make clear why animals are not born with a memory of how to do something that they have never done before.

First, as observed by Weismann, instinctive behaviours probably evolved out of reflexes and can still be released by stimuli, just like their precursors. Hence instincts are inborn dispositions to react. Second, animals do not *recollect* 'phylogenetically stored knowledge' (as Plato misguidedly tried to argue in the *Meno* in order to prove the immortality of the soul). For the behaviour exhibited by animals does not license us to say that animals *use* this alleged 'stored knowledge' for their responses or for developing goal-directed behaviour. Their instinctive behaviour lacks the flexibility typical for behaviour based on acquired knowledge. The idea that instincts may involve stored knowledge is also based on the mistaken presupposition that knowledge is *stored*, for there is no such thing as storing knowledge (in the mind, brain or DNA). I will elaborate this point below when I discuss the second homunculus fallacy. Third, as explained above, instinctive behaviour is displayed unhesitatingly, but this does not mean that knowledge is involved, for certainty, doubt, mistake, misidentification or misrecognition are all logically ruled out. Hence the ability to behave instinctually is not an example of a two-way ability (or power), for animals cannot refrain from behaving instinctually. If animals, after they have learnt something, *use* this (recollected) knowledge, then they are able to refrain from doing something. In such cases it makes sense to say that they are able to choose and can behave at will. Instinctual actions and reactions, however, are behaviours in which memory does not serve. Fourth, as Russell noted in his *Problems of philosophy* (1967 [1912], p. 49), 'a main objection which seems fatal to any attempt to deal with the problem of innate knowledge' or representations, is that our nature would change as the result of mutations. Might it happen that 'our nature would so change as to make two and two become five'? Russell discussed a mathematical truth, but similar problems arise if one argues that conceptual truths (for example, 'red is a colour') are examples of innate representations or knowledge. For it is misleading to speak about an innate representation in the brain (or in the DNA) if it is meant to signify a symbolic (or semantic) representation. Similar problems arise when it is argued that there are brain or genetic systems that *operate* upon representations (see Hacker, 1990a). For genes or brains are not sentient beings that can be said to employ symbols that can be used to mean something. Genes and brains cannot correct mistakes, cannot explain why it is mistaken to say that two and two is five; that saying 'red is a triangle' or that pointing at a tomato and saying 'this (☞) is blue' is mistaken, etc. Conceptual truths, i.e. the expressions of the internal relations of a symbolism, are part of the conceptual scheme or framework that we use as language-using creatures, but they do not belong to our biological nature. Of course, constancies and regularities in human nature and nature are a prerequisite for our symbolism. The rules of tennis would be different if we could play tennis on the moon, and our concepts of colour would be different if the light of the sun was infrared or if we had the facet eyes of insects. But our certainty (the absence of doubt) that two and two is four, that red is a colour or that a rod has a length, is not accounted for by referring to innate knowledge, representations or concepts. The apparent harmony between knowledge and the world (and language and the world) is not grounded in human nature.

This conceptual analysis enables us to resolve the problem mentioned in the previous sections: since Gottlieb and others have shown that prenatal experience (self-stimulation) contributes to the development of the discrimination abilities of ducklings, does this mean that the development of this ability is no longer an example of an instinctual response since it involves learning? Based on the analysis given above we can conclude that prenatal experiences do not result in *cognition*. There is nothing in the behaviour of the ducks that licenses us to say that the ducks *remember* these experiences or that they can *use* these retained experiences for future behaviour. Self-stimulation only accelerates the development of the discrimination ability, i.e. the ability to respond to the maternal call. The contribution of this experience to the development of the ability may have evolved since it increased the survival rates of hatchlings. But Gottlieb's investigations do not license us to say that the animals remember anything after self-stimulation, for they do not display behaviour that shows that they sometimes forget something, are ignorant or recollect what they have forgotten. Just as habituation, sensitization and conditioning are not examples of learning processes that warrant the ascription of memory to an animal, self-stimulation is not an example of information retention. Hence learning something through self-stimulation, habituation, sensitization or conditioning does not involve cognition and does not involve the use of knowledge retained. These learning processes lead only to changes in the reactions of animals or in the formation of dispositions to behave.

6.7 Behaviour, brain and mind

Weismann illustrated his objections against the theory of Lamarck with the example of an actor. According to him it is for biological reasons implausible to suggest that alterations of memory cells in the brain (as the result of memory-exercises of the actor) cause adaptive changes in the determinants present in the germ cells. Since Weismann's example involves knowledge, it invites the conceptual problem of how we can investigate the role of 'memory cells' in the retention of knowledge. For an actor does not learn something through conditioning, but acquires knowledge and is able to use this knowledge. After he has learnt to play his role he knows how to perform on stage. Is this knowledge then *stored* in brain cells or structures and do these putative memory cells or structures *contain* knowledge after the memory-exercise? Wittgenstein has argued that there is no knowledge stored in the brain. For if one says that information is stored in the brain, then one commits what Kenny (1984, chapter 9) later called a homunculus fallacy. Wittgenstein formulated his conceptual insight as follows:

> Only of a human being and what resembles (behaves like) a living human being can one say: it has sensations; it sees; is blind; hears; is deaf; is conscious or unconscious. (2009 [1953], par. 289)

The fallacy discussed by Wittgenstein is ascribing to a part of a creature that which logically can be ascribed only to the creature as a whole. So it makes sense to say that animals and humans can, for example, see, since they, as living, sentient beings, use their senses and, hence, have the faculty of sight. But it is senseless to say that cells or parts of the brain can see, since cells or brain parts do not use senses. If someone says, for example, that a cell or a brain sees, and intends to investigate how these parts of a human being see, then we would not know what kind of evidence he can present. For cells and brains do not use senses and hence cannot inspect, observe, notice, remark something, etc. There are several kinds of activities in the brain but cells and parts of the brain do not use senses and do not form intentions.

It was already noted by Kenny (1984) that the term 'homunculus fallacy' may be misleading. First, committing the fallacy is not simply ascribing a psychological predicate (like seeing or remembering) to an alleged homunculus in the head. For although the existence of homunculi in the sperm cells was postulated by scientists in the past, investigators do not postulate the presence of a homunculus in the skull. The error in question is ascribing psychological predicates to parts of the animal that apply only to the whole (behaving) animal. Bennett and Hacker (2003, 2008) prefer for that reason the term 'mereological fallacy' (mereology is the logic of part/whole relations). Yet Kenny had, of course, a reason for using the term 'homunculus fallacy', for if someone ascribes psychological predicates like 'seeing' or 'believing' to the brain, this invites the question how the brain can see and believe (and asking this question is tantamount to asking whether there is a homunculus). Second, the error is not an example of invalid reasoning, but, as Kenny pointed out, it leads to fallacies. If someone ascribes psychological predicates to the brain, then this introduces incoherence in his conceptual framework that may result in invalid reasoning.

The error noted by Wittgenstein may be seen as a conceptual extension of Weismann's critique on the ideas of Lamarck. The error discussed by Weismann is that it is unlikely that there are modifications in the germ cells corresponding to the functional adaptations in brain cells, since the determinants in germ cells are not miniature memory cells. Hence alterations in memory cells, which are a prerequisite for retaining information, are different from modifications occurring in germ cells. The error discussed by Wittgenstein a not an empirical but conceptual one: although changes in brain cells are a prerequisite for memory, there are no memory cells in the sense that these cells retain information. It makes sense to say that an actor has a memory, is able to retain information and possesses knowledge; yet it makes no sense to say that the hippocampus or cortex possesses knowledge or that information is stored in these brain areas. It is the whole animal that retains information and exercises the ability to retain information, not its parts. Why are memories not stored in the brain? Why can facts of consciousness, as Spalding noted, not be thought of as packed away like books in a library?

Memory is knowledge retained, but this knowledge is not stored in the brain. The idea that knowledge antecedently acquired is stored in the brain is based on the false (empiricist) presupposition that what is remembered must be available to one. Hence when someone remembers something, it is believed that a current experience somehow activates an antecedent experience that is present as a trace or representation in the brain/mind. The problem is that there is no stored representation that is available to one, for that presupposes that we are able to observe or read the putative representation in the brain/mind (as we can observe a picture stored in an album or a remark in a diary). But this is incoherent since there is no such thing as storage of representations in the brain/mind. To remember something is to possess knowledge but not to store it (one can store information in a diary or a picture in an album, but one cannot store information in the brain). Of course, it is possible that neural processes (determining, for example, the strength of synaptic connections) or, less likely, genetic recombination in brain cells (see Pena de Ortiz and Arshavsky, 2001) are a prerequisite for remembering a certain piece of information. But it does not follow that these neural processes are representations of this information. On the contrary, there are no representations encoded in the brain/mind.

The thought that memories are stored in the brain results in misleading ideas about the role of brain areas in memory. It also invites the (science fiction) problem of whether memories can be transferred from one individual to another (just as Balaban, 1997, 'transferred' components of instinctive behaviours through implanting parts of a quail midbrain into chick embryos). Is it possible to transfer memories from one individual to another? Can we, as Spalding asked his readers in 1873, create a human being with the same personal identity as our friend, through forming by a mysterious process someone who is the same in every particle and fibre as our friend (note that Spalding was not referring to identical twins)?

First, the particular thoughts and memories I have during later developmental stages are my experiences. What I as an individual am thinking or remember is, therefore, dependent upon me. There are, of course, alterations in brain processes when I experience something. These are causal conditions for my experiences but are not these experiences. Second, these experiences cannot be transferred from one human being to another since they are not chemical substances like genes. Experiences are expressed in our behaviour, i.e. they are, in Aristotelian jargon, characterized by the form of a substance. But just as we cannot, for logical reasons, investigate the idea of alchemists that the quality of a chemical element is transferable from one element to another, we cannot investigate experiences severed from the brain (see Hacker, 2007, chapter 10). The experiences I have belong to me but cannot be separated from me, just as the quality of an element belongs to it. Of course, we can transform elements and chemical molecules, just as we can transform the genome of a human being. Transformations in the case of humans will change universal traits (and, in the end, particular experiences in so far as universal characteristics are constitutive for later experiences). But that does not prove that memories can be transferred. Third, a particular experience is not analysable in terms of a combination between states of the mind and states of the brain (as Spalding thought). For if we think so, we assume an incoherent (Cartesian) dualism between mind and brain. Dualism was in the past a source for science fiction (see Wiggins, 2001, chapter 7). It was the reason why the empiricist Locke mistakenly thought that the identity of a person lies in mnemonic continuity. Of course, Locke thought that past experiences inhere in something, but he believed that consciousness of past experiences does not require that the person conscious of his past experiences inheres in the *same* body (hence the science fiction about transferrable memories). This idea of 'psychological continuity' as necessary for someone's identity leads to the absurd idea that an amnesiac would be a different person. Of course, someone may have little sense of his own identity if he, because of an accident, no longer possesses knowledge of what he did in the past. But it does not follow that he is a different person from the person before the accident.

6.8   Conclusion

The combination of Weismann's and Wittgenstein's ideas clarifies our understanding of the harmony between instincts, knowledge and the world. Weismann has argued that the ideas of Lamarck are false since Lamarck's theory tacitly presupposes that the determinants in the germ cells are miniature memory cells. It is therefore, according to Weismann, unlikely that there are corresponding effects in the brain and germ cells as the result of the memory-exercises of an actor. The upshot of Weismann's critique is that instinctive behaviours do not evolve as the result of habit in use or disuse, but are selected during the course of evolution. Wittgenstein has explained why it is for conceptual reasons mistaken to believe that there are memories or ideas stored in the brain or in the genome. Instincts do not involve knowledge, for animals do not recollect phylogenetically stored ideas in their brains or DNA. Thinking that innate knowledge brings about the harmony between an organism and its environment is committing a conceptual, homunculus fallacy, for there is no such thing as storing knowledge in the brain or in the genome. Whether an animal knows what it does depends on what the animal can do. Instinctive behaviour is at first displayed unhesitatingly and is species-specific, but may be modified or controlled by animals when they start using their senses, become self-moving creatures starting to explore the world and when their intellect and will develops as the result of maturation and learning. We have to observe their behaviour in order to determine whether an animal uses acquired knowledge for pursuing self-chosen goals. If an animal uses knowledge, then it is able to discern errors and to readjust its goals. While humans, as language-using creatures, can explain mistakes and tell of changes in their intentions, it is for obvious reasons harder to investigate whether animals use 'primitive knowledge' for developing plans and preferences.

The ideas of Weismann and Wittgenstein explain why instinctive behaviour precedes the development of reasoned behaviour. The reason is simple: there is no such thing as the genetic transmission of knowledge from one generation to the other. Yet the ability to learn something has, of course, been subject to selection, for if there is genetic variation affecting the ability to learn something and if this variation affects reproductive fitness, then there will be selection on genotypes underlying the ability to acquire knowledge. Genotypes determine in interaction with the environment the development of phenotypes. And phenotypes interact with the environment, that is, there is selection on phenotypes with the result that some genotypes survive. But this does not mean that genotypes or phenotypes (brains) contain information. An investigation of two homunculus fallacies teaches us that instincts and cognitions are not stored anywhere.

CHAPTER 7

Language evolution: doing things with words versus translating thought into language

7.1 Introduction

Evolutionary theorists have recently discussed the possibility that our knowledge of language is innate and explicable as an evolutionary adaptation (see, for example, Tooby and Cosmides, 1992; Maynard Smith and Szathmáry, 1995, 1999; Pinker, 1994; Scott-Phillips, 2010; Szamado and Szathmary, 2006; Hurford, 2007; 2012). Some of them have used Chomsky's ideas as a model and argue that humans are equipped with an innate language faculty. These evolutionary theorists assume that future research will provide us with more insight into the nature of this faculty.

The language faculty that humans possess is described as a system that is present in the mind/brain. This system is thought to consist of an abstract Universal Grammar incorporating a set of principles. Some of these principles are fixed and rooted in our genetic endowment; others are characterized as parameters which obtain a value as the result of learning. This learning process is, however, according to Chomskyans not comparable to how psychologists normally conceive of learning processes. They argue that an innate Acquisition Device somehow selects relevant data appropriate for the development of a language. This selection process is seen as the core of the 'learning' process. The result is that the parameters appropriate for the environment in which the child is living are, as it were, switched on, resulting in the ability to speak and understand a language. One argument why proponents of Chomsky's model argue that traditional learning theories are misguided, is that they believe that children learn their native language fast. Another argument is that studying theories about language does not improve the capacity to speak and understand a language.

In the first part of this chapter I shall discuss Chomsky's model further and ask whether this model (and related cognitive models) is suitable for investigating language evolution. I shall argue that it is not because the model is conceptually incoherent and at variance with models developed in evolutionary biology. Since Chomsky had presented his model as a solution to problems discussed by philosophers, I shall also discuss the question of whether his model resolves philosophical problems.

In the second part of the chapter I shall discuss an alternative model. This model is based on the observation that language enables us to do things with words, i.e. it enables us to engage in new forms of communicative behaviour. This observation raises the possibility that language evolution is explicable in terms of inclusive fitness theory. I shall explore this possibility through investigating how linguistic behaviours evolved as extensions of behaviours and gestures which were displayed by the early hominids. This investigation clarifies how language evolved subsequent upon bipedal locomotion, because bipedal locomotion led to new functions of the hands. Tool-making, pointing and natural pedagogy, co-evolving with changes in attention, imitation and social cognition, led to the first use of words. The use of words in simple communicative behaviours was later extended with more complex ones.

7.2   Selection versus instruction

Chomsky's ideas are not based on empirical data nor deduced from a theory. It is therefore remarkable that some evolutionary biologists have treated his ideas as if they are part of a scientific model. Why have they accepted his ideas uncritically? One reason is that Chomsky has advanced his theory as a critique of, and alternative for, behaviourist learning theories (Chomsky, 1959). Chomsky (who considers himself a Cartesian linguist) developed his ideas at a time when some behaviourists believed that studying the genetics of verbal behaviour was irrelevant for understanding the ontogenesis of the ability to speak and understand a language. These behaviourists neglected the role of internal factors (i.e. the maturation of the brain affected by expressed genes) and studied verbal behaviour as the result of external factors (stimulation and reinforcement). Because ethologists and others have shown that internal factors contribute to the development of behaviour, and since geneticists have shown that genes are involved in certain learning processes, there are good reasons for evolutionary biologists and psychologists to reject behaviourist 'blank slate' theories (Pinker, 2002; Smit, 1989). A second, related reason is that Chomsky has discussed the problem of how children acquire a language in terms of the opposition between instruction and selection. The opposition between behaviourist learning theories and Chomsky's own theory is therefore given a semblance of well-known oppositions in life science. Because selection theories have superseded instruction theories in life science, the successes of selection theories are seen as an argument in favour of Chomsky's ideas. However, this argument is flawed and it needs a brief discussion of these theories to explain why.

Some of the major breakthroughs in life science in the previous century can be described in terms of the general opposition between instruction and selection (see, among others, Jerne, 1967; Lederberg, 2002). I shall discuss three examples. First, Delbrück and Luria (1943) have shown with the help of their so-called fluctuation test that the resistance that bacteria develop against bacterial viruses does not evolve as the result of an instruction process (as was thought by others), but is the outcome of Darwinian selection on genetic variation in populations. Second, Jacob and Monod (1961) discovered that the ability of bacteria to produce enzymes necessary to digest lactose does not arise through an instruction process, but results from the activation of genes already present. When lactose is added to the environment, the repressor (which is normally present) is 'removed' from the promoter site of the protein synthesis machinery. Evolutionary theory explains why the gene involved in the production of lactase is only expressed if lactose is present in the environment: if the signal is absent, expressing lactase imposes a cost and, hence, does not improve survival and reproduction. Third, Jerne (1955) and Burnet (1957) have argued that the ability of the immune system to recognize antigens and to respond to an enormous diversity of foreign antigens arises as the result of selection of antibodies generated by genetic recombination and somatic mutation. Their theory superseded the older instruction theory (Breinl and Haurowitz, 1930; Pauling, 1940) which stated that organisms 'learn' to recognize antigens: the antigen acts according to this theory as a template during protein synthesis, moulding the final shape of an antibody (with the result that the antibody fits the specific antigen). Burnet, Jerne and others demonstrated the falsity of the instruction theory. In Burnet's clonal selection theory, cells that are capable of producing an antibody that can bind the antigen are selected and start to proliferate after exposure to the antigen. Since the response of organisms to the same antigen is faster during a second exposure, it is said that the immune system creates an adaptive immunological memory.

These three examples can all be understood in terms of the general opposition between instruction and selection. But notice that there are interesting differences between them requiring evolutionary explanations. The experiment of Delbrück and Luria illustrates Darwinian selection at the level of unicellular creatures and demonstrates the falsity of Lamarck's theory. The experiments of Jacob and Monod show that bacteria have an innate, first-order capacity to adapt themselves to changes in the environment, but this ability is only visible if it is 'triggered' by a substance in the environment. The development of an adaptive immunological response is the result of selection of cells during the lifetime of an individual, and is an example of an acquired, second-order ability made possible by specific genetic and cellular mechanisms. Hence we can describe these examples in terms of the opposition between instruction and selection, but they also show that during evolution first- and second-order abilities evolved enabling cells and later organisms to respond to challenges from the environment. I will return to this latter observation when I discuss Chomsky's observations on the use of the term 'ability'.

Humans have evolved a hierarchy of responses which enable them to cope with challenges over a number of timescales (Gluckman and Hanson, 2004; Gluckman *et al.*, 2009; see Figure 7.1). Some of the responses are rapid (mediated by the nervous and endocrine systems) and involve homeostatic mechanisms, others involve within-lifetime adaptations mediated by epigenetic mechanisms (so-called predictive adaptive responses, i.e. responses made for anticipated need or advantage later in the life course). Yet another response is beyond the timescale of individual lifetimes and involves natural selection resulting in genetic change over several generations. Evolutionary biologists have developed detailed evolutionary models explaining how and why first- and second-order abilities evolved. In these models proximate and ultimate explanations are integrated into a coherent model. Chomsky's investigations have never led to a research programme demonstrating how explanations of linguistic behaviour can be integrated with knowledge developed in (evolutionary) genetics, neurobiology and psychology. One reason is that Chomsky has from the start insulated his theory from knowledge developed within other fields in science. He has argued that evolutionary theory is not useful for understanding the evolution of the language faculty, since humans *suddenly* acquired the capacity to use a language (and this observation is, according to Chomsky, at variance with the theory of evolution). Although Chomsky later modified his view (see Hauser, Chomsky and Fitch, 2002), attempts to integrate his theory with models from evolutionary biology have not been successful. Similar problems arise with regard to psychology and neurobiology. According to Chomsky, the cognitive system in the mind/brain possesses 'knowledge', yet this form of knowing is different from how we normally conceive of knowing. The 'knowledge' that the

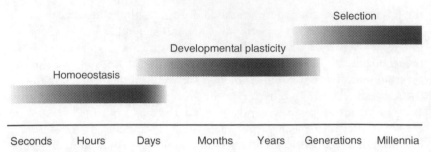

Figure 7.1: Responses over different timescales (adapted from Gluckman *et al.* 2009).

mind/brain possesses is inaccessible to current neurological investigation. This assumption makes it difficult to integrate his model with knowledge from psychology and neurobiology. Hence my observation: although Chomsky's theory has a semblance of selection theories in life science, it does not follow that Chomsky's theory is a suitable model for studying the evolution and ontogenesis of the ability to speak and understand language. It raises the conceptual problem of why Chomsky's model is at variance with knowledge developed in adjoining fields.

7.3   Innate knowledge

Chomsky assumes that children have an innate Language Acquisition Device (LAD). This is conceived as a 'neural mechanism' that causes a child instinctively to acquire and follow grammatical rules on the basis of cues or data provided by its linguistic environment. He assumes that the LAD is present in the mind/brain and enables children to develop the ability to speak and understand a language. The innate language faculty is described by Chomsky as a system incorporating grammatical principles. These principles are innately 'known' by the mind/brain but not by the child: Chomsky assumes that the faculty is at first present at a subconscious level. Yet this system 'grows', according to Chomsky, in the mind/brain, and once it has grown, the child is able to understand and speak a language. But what is meant here by 'growing' is obscure because Chomsky talks about a system present in the mind/brain. The problem is that the brain consists of cells, but the mind not. Some things can grow in the brain (e.g. a tumour) but not in the mind. And if something grows in the mind, for example, the suspicion that Chomsky's ideas about systems 'growing in the mind/brain' are flawed, then this suspicion does not grow in the brain. The idea that the ability to understand a language grows in the mind/brain is therefore conceptually incoherent.

In his well-known principles and parameters model, Chomsky assumes that the language faculty or instinct consists of Universal Grammar which is present as an analytic device in the mind/brain. We should, according to Chomsky, think of the language faculty as a complex and intricate network of some sort, associated with a switch box consisting of an array of switches that can be in one or two positions (Chomsky, 1988, p. 62). Language learning is the process of determining the values of the parameters left unspecified by Universal Grammar. Chomsky insists that 'learning is not really something that the child does; it is something that happens to the child' (Chomsky, 1988, p. 134). When children learn a language, their mind/brain 'interprets' the incoming linguistic data through the devices provided by Universal Grammar. Chomsky argues that this is not an instruction but a selection process: 'The language faculty selects relevant data from events taking place in the environment; making use of these data in a manner determined by its internal structure, the child constructs a language' (Chomsky, 1988, p. 35). He assumes that Universal Grammar is, in terms of the computer metaphor, hardwired, and through parameter setting, children acquire rules belonging to the software of the mind/brain. This explains, according to Chomsky, why only humans acquire a language: they possess innate knowledge (Universal Grammar) and this knowledge is, through parameter setting, transformed into a specific language (English, Dutch, etc.).

However, Chomsky's ideas are at variance with those generally accepted in life science. When biologists talk about faculties or abilities, they do not refer to 'cognitive systems'. They make a distinction between first- and second-order abilities. A first-order ability is, for example, the ability of an organ to exercise its function, for example, the ability of the kidney to process waste products. The development of this ability does not require learning and teaching (organs must, of course, mature) and is in this sense innate. The language faculty is not an example of a first-order ability but of a second-order ability: humans possess an innate ability to acquire a language. Children may or may not learn a language fast (see further below), but the ability to acquire a faculty is, in part, determined by internal factors, i.e. there is a critical period. This critical period is explicable in terms of the growth and maturation of (certain areas in) the brain and in terms of processes such as synaptic plasticity and pruning. The second-order ability explains why only humans can understand and use grammatical rules: other animals lack the innate ability to acquire a language.

Chomsky contends against these well-known distinctions in life science. He argues that the language faculty is a component of the mind/brain and it is, according to Chomsky (1988, p. 55), a matter of 'biological necessity' that children possess an innate system (consisting of rules) in their mind/ brain. What is meant here by 'biological necessity' is, however, left unexplained. Empirical data showing that children are endowed with an innate system consisting of abstract rules (which they do not know) are not given (and are not available). Chomsky's remark, that the rules are hidden from consciousness, does not help here and only mystifies problems. It also does not help to say that the language faculty in the mind/brain 'processes data' and that this 'processing' is introspectively inaccessible to the conscious mind, for the brain does not process data. And these conceptual problems are not resolved by saying that children do not know but 'cognize' (Chomsky, 1980, p. 69) the principles of Universal Grammar, as long as no precise definition is given of the allegedly scientific term 'cognize'.

7.4   Aphasia

Chomsky has objected to the idea that humans have an innate ability to acquire a language. It is important to recall that Chomsky developed his ideas as a critique of theories of behaviourists about acquiring verbal behaviour as the result of stimulation and reinforcement. The ability to acquire a language was explained by some behaviourists in terms of dispositions and habits 'caused' by external factors. However, as I have explained above, the term 'ability' is used in life science in a different way: it refers to first- and second-order abilities which arise, in part, as the result of internal factors. The critique of Chomskyans on blank slate theories is, therefore, not applicable to theories from life science. Of course, there are differences between the innate ability of the immune system to acquire immunity and the innate ability to acquire a language, for only the ability to speak and understand a language involves knowledge. Hence the problem arises whether this difference justifies Chomsky's argument that we should no longer speak about an ability in the case of language. I shall investigate this problem further through discussing the example of aphasia.

In his critique of Chomsky's ideas, Kenny (1984, chapter 10) suggested that knowledge of language is akin to an ability and used aphasia as an illustration. In a response to this critique Chomsky (1988, pp. 9ff.) elaborated his theory and advanced the idea that aphasia is better explained in terms of his theory. Chomsky noticed that aphasia shows why the concept of ability is of no use here and should be replaced by his scientific terms (the cognitive system in the mind/brain). His main argument is based on a medical observation: there are cases of aphasics who lost the ability to speak and understand a language after brain injury, but recovered and were able to speak and understand this language again without any training or learning. According to medical doctors, recovery from aphasia is the *restoring* of an ability, just as recovery from blindness is the restoring of the faculty of sight. Chomsky, however, believes that these medical ideas are mistaken. Recovering from aphasia shows, according to him, that the cognitive language system (in the mind/brain) remained intact when patients suffered aphasia.

In order to understand the medical explanation, it is useful to make some conceptual distinctions (see Kenny, 1984; Hacker, 1990a). First, we can distinguish an ability from how it is exercised. Humans have all kinds of abilities, for example, the ability to play tennis, to speak a language or to do calculations. We ascribe an ability based on what he or she does, just as we ascribe chemical powers to substances based on the reactions they cause. The more complex the ability, the more diverse and diffuse these grounds are. Yet abilities are only visible if exercised, and ascribing an ability to someone depends therefore on *opportunity conditions*. If there are no opportunity conditions, then one cannot display one's abilities (for example, there is no tennis court available). Second, we can also distinguish external from internal constraints (distress as opposed to someone being ill or drunk) on the exercise of an ability. Suppose that someone is very drunk and unable to speak and understand English, i.e. the alcohol constrains the exercise of his abilities. Is an aphasic comparable to the drunk? Medical doctors would say that he is not, for aphasia is not a *constraint* on the exercise of an ability but, like blindness or paralysis, a *loss* of an ability. According to them recovering from aphasia is explicable in terms of neurophysiology.

Since aphasia involves knowledge, we have to answer Chomsky's question why biologists and medical doctors still use the term 'ability'. They use the term 'ability' since we determine whether someone knows something by observing *what he or she can do*. If someone possesses knowledge, then he or she has the ability to answer several questions, is able to explain something or can show how it is done. And whether someone remembers something is determined by observing his ability to retain what he came to know, i.e. whether he can retell and explain something or can still exercise an acquired ability. To forget something is ceasing to be able to do something. Is aphasia comparable to forgetting something? If I forget a fact (the occasional lapses of memory most of us have), then someone may remind me of this fact. Or someone may give me a hint so that I recall what I had forgotten. In these cases we recollect knowledge that we have forgotten. They show that our memory is not always reliable. Aphasia is, of course, not comparable to forgetting something, for it is conceptually incoherent to ask whether we can *remind* an aphasic of something that he has forgotten. Is aphasia comparable to amnesia? In the case of amnesia someone suffers from loss of memory. Sometimes someone regains his memory but this regaining is not comparable to recollecting something that one has forgotten. For we cannot remind the amnesiac what he has forgotten, and when his memory is suddenly restored, it is misguided to say that he had suffered previously to that from a lapse of memory. Recovering from amnesia is, according to biomedical investigators, regaining memory. Aphasia is, according to them, not comparable to amnesia, for recovering from aphasia is restoring the ability to speak and understand a language.

Chomsky objects that there is a difference between knowing a language and being able to speak and understand it. According to him, an aphasic possesses knowledge although he is unable to speak and understand a language. The recovery from aphasia shows, according to Chomsky, that this knowledge is *retained* (Chomsky, 1988, p. 10). And since he has not learnt the language de novo, the conclusion, he suggests, must be that he has retained knowledge somewhere in his mind/brain. If I were to stop here, I guess that some readers would prefer Chomsky's ideas to those of his opponents. For how else can we explain how a Dutch aphasic speaks his native language again after recovery? As Chomsky (1988, p. 10) put it: 'Plainly, something was retained.' Doubts, however, arise as soon as we ask how we can test Chomsky's ideas. Recall that Chomsky objects to the use of the term 'ability'. He prefers the term 'state' as a sound alternative. There is, according to Chomsky, a state of the mind/brain which remains present while patients suffer aphasia. Hence we can ask whether we can develop testable hypotheses about this alleged state of the mind/brain. Is this state of the mind/brain explicable in terms of genetic, neurophysiological or evolutionary knowledge? The answer is 'no' and there is a simple reason. If one assumes that knowledge is somehow and somewhere retained, then it is presupposed this knowledge is present in the brain. Yet the problem is that the brain is not a storehouse of knowledge. There are all kinds of processes in the brain (action potentials, etc.), but there are no symbolic representations encoded in these processes. Chomsky's ideas lead therefore to interminable debates about where and how knowledge is stored (see also Bennett and Hacker, 2008, chapters 3 and 4). Notice the reason why Chomsky's ideas end with conceptual problems. His ideas are generated by the replacement of the term 'ability' by the term 'state'. This conceptual shift mistakenly reinforces the impression that there is an *empirical* problem involved (Hacker, 1990a). For believing that knowing is a state results in the mistaken idea that knowing something is to be in a *persistent state*, i.e. it leads to the idea that a piece of information is stored somewhere in the mind/brain so that it is available for future use. It also leads to the thought that knowledge is deposited in the memory banks of the mind/brain. And since it results in the belief that we are dealing here with empirical problems, it strengthens also the impression that we can find a causal link between past learning, retaining knowledge and recollecting knowledge upon recovery from aphasia. In the end Chomskyans are perplexed. Why is it so difficult to study the alleged states of the mind/brain where knowledge is stored? Why is it so hard to integrate the model with knowledge developed in genetics, neurobiology and evolutionary theory? The answer is simple: because the model is conceptually incoherent.

The biomedical explanation of aphasia is not confronted with these conceptual problems. In life science, investigators prefer to use the term 'ability' rather than the term 'state'. Abilities cannot be located in space (this characteristic of abilities explains, perhaps, Chomsky's objections to the use of this term). For example, if we want to explain the ability of the immune system to respond to an antigen, we refer to the structure of antibodies and to certain cellular mechanisms (just as we can explain the ability of a key to unlock this door by referring to the shank of the key or to the keyhole in the door in which the key exhibits its powers when turned). We are then referring to the vehicle of a power or ability. Antibodies are Y-shaped and the two arms of the Y possess so-called hyper-variable parts that differentiate one antibody from another. These parts are generated by special genetic mechanisms (somatic recombination and mutation) and this explains why the immune system has, as it were, millions of keys that fit the different locks (antigens) present in the world of pathogens. This explanation of abilities and their vehicles clarifies why alterations of the chemical structure of antibodies affects their ability to bind and respond to antigens (hence we can transform an ability). It also clarifies why we cannot transfer a chemical ability or power from one substance to another. And it also makes clear why immunologists do not use the picture of 'storing knowledge' when they talk about the immunological memory (but talk about the second-order ability of the immune system to acquire immunity). If someone is not able to develop an immune response to a certain antigen and later spontaneously develops this ability, then we do not say that the immune system forgot something and later recollects what it had forgotten (empirical studies have shown that there is then a mutation involved altering the course of the immune response, see for an example, Hirschhorn *et al.*, 1996). It is the conceptual picture of storing knowledge that mistakenly reinforces the misguided thought that the immune system has suddenly found the knowledge necessary for the response. Similarly, when an aphasic recovers, the picture of knowledge being somewhere stored in the mind/brain generates the illusion that stored knowledge is somehow retrieved. But regaining an ability is only misguidedly interpreted in terms of 'finding knowledge in the mind/brain' if our picture of storing knowledge is used.

Chomsky not only believes that there is a cognitive system in the mind/brain, but also that using this system is akin to using a theory. The 'fact' that children acquire grammars of great complexity suggests to him that the alleged language faculty is 'designed' as a hypothesis-forming system. Otherwise, according to Chomsky, it is inexplicable why children are able to understand new sentences. Again, these conceptual problems do not arise if we use the term 'ability'. Some abilities (for example, instinctive responses) are rigid; other abilities are plastic. If knowledge is involved, as is the case in mastery of language, then the ability to use a language is open-ended (see Baker and Hacker, 1984; Fischer, 1997; 2003). In contrast to instinctive behaviours, no single act or activity is then a criterion for the exercise of an ability. Hence it is unsurprising that someone who has mastered a language is able to understand many sentences, including new sentences that he or she has not heard before. Chomsky's suggestion that understanding a language is comparable to knowing a theory does not help here (and only creates conceptual problems). For there are obvious differences between knowing a theory (say the theory of evolution) and knowing a language. A child learning English does not learn facts confirming linguistic hypotheses or models about this language. A language is not a scientific theory, and it makes no sense to talk about the validity of a part of language (while we can doubt the validity of parts of a theory). For what would a false language be? Someone who knows a language has not mastered a theory but is able to answer questions such as what words mean, what correctly formulated sentences are and what sentences mean.

7.5   Is the Baldwin effect a solution?

If the language faculty is not a first-order ability but a second-order ability, then it is possible that natural selection has shaped this ability. Some evolutionary biologists have argued that the Baldwin effect (also named the Spalding effect; see Chapter 6; Spalding, 1954 [1873]) may be helpful here. They have argued that applying this effect to language evolution enables us to reconcile Chomsky's theory and the theory of evolution.

Suppose that the (already present) ability to acquire a language is subject to adaptive evolution. It is then possible that fast learners have an advantage compared to slow learners. Baldwin (1896) argued that adaptive behaviours can evolve through learning processes if these behaviours require a neural 'rewiring' of the brain. Assume that there are genes affecting the 'rewiring' process and that adaptive evolution is involved. Then it is possible, according to Baldwin, that once learnt behaviour becomes instinctive. Applied to Chomsky problems, it has been argued that the LAD may have been evolved through this process (Glackin, 2010). Initially language was acquired through learning, but since there was a selective advantage accruing from a reduced learning cost, less 'softwiring' of parts of the mind/brain had a selective advantage with the result that some parts became 'hardwired'. In the end, Universal Grammar evolved as an instinctive, 'hardwired' part of the mind/brain. In terms of the principles and parameters model it is argued that the principles evolved as the result of the Baldwin effect, while the parameters are the subset of the rules which are not fixed but obtain a value as the result of 'learning'.

In this model it is assumed that genes affecting the neural circuits 'take over' a large part of the learning process. The result may be that, eventually, so little learning is needed that it comes close to being an innate or instinctive ability. It is, however, not an instinct, since someone who uses a language responds differently to the correct and incorrect use of rules (i.e. he or she understands why it is incorrect to say 'John Mary loves').

Does this reformulation of Chomsky's theory in terms of the Baldwin effect improve Chomsky's core ideas? As long as we use the Spalding or Baldwin effect to describe the possibility that genes affect the ability to acquire a language, there are no conceptual problems (see Chapter 6). However, if it is assumed that this effect explains why Chomsky correctly argued that there is a 'hardwired' LAD present in the mind/brain, then we are confronted again with problems. For this reformulation invites the empirical question whether we can study this device as a state of mind/brain, and creates therefore the conceptual problems discussed above.

7.6 The evolution of doing things with words

I turn now to the alternative model. I shall sketch a scenario of language evolution based on both empirical and conceptual insights, and argue that language did not evolve as the result of mutations affecting (nonconscious) knowledge of a grammar or lexicon in the mind/brain of hominids, as Chomsky and others assume, but shall elaborate the hypothesis that language evolved because it enabled hominids *to do things with words* (see Wittgenstein, 2009 [1953]; Baker and Hacker, 2005 [1983]). To understand language evolution is to explain how linguistic behaviour emerged as a new form of communicative behaviour. The aim of the scenario is not to tell a once-and-for-all true story, but to describe the sequence of the early shifts and transitions in human behaviour (and social cognition) that led to language evolution. I hope that the scenario inspires future research aimed at developing hypotheses which can be tested through comparative, developmental and genetic studies.

Children have an innate ability to acquire a language. When they learn to use words and later sentences, they learn to do things with words (together with gestures) and acquire then the ability to participate in typical human activities. To acquire a language is to learn new forms of action and activities and forms of reaction and responses to speech. These linguistic actions and responses are constitutive to a human form of life. Examples are learning to demand, beg and request; to call people and to respond to calls; to express needs, sensations and emotions and to respond to the expressions of others; to ask and answer questions; to name things and to describe and to respond to descriptions of how things are, and so on and so forth. These are forms of communicative behaviour, but the important point to notice is that they are *not* examples of communicating thought (asking where Peter is, is not communicating one's thought; telling one's intentions is not telling what one thinks; and threatening and warning are not a matter of expressing thoughts). Humans can, of course, communicate thought, but this is something that evolved during later stages of our evolutionary history because it is an example of a complex skill. For example, when someone says: 'I thought that Peter would come tonight, but later noted that he changed his plans', then this expression of his thought presupposes that he can use a tensed language, understand what intentions and plans are, understand that someone can change his plans, etc. Hence to solve the problem of how language evolved is not to answer the question of how, during the course of evolution, thought (concepts, ideas) was translated into the medium of language for purposes of communication (as is assumed by Chomsky and others, see Chomsky, 1988, pp. 27ff.), but to answer the question of how new forms of communicative acts and responses (like demanding, asking, ordering) evolved as an extension and replacement of behaviours already displayed by our predecessors. It also concerns answering the question of how the 'simple' linguistic behavioural repertoire was later extended with more 'complex' communicative acts like expressions of intending, thinking and imagining. This distinction between simple and complex forms of communicative behaviour parallels the evolution of the use of complex tools and the first forms of art (about 150,000 years ago) out of the use of simple tools (2 million to 3 million years ago). For art (e.g. representational art) presupposes abilities like imagining and, hence, the ability to use more complex linguistic skills (see Chapter 2, section 2.2).

The answer to the question of why language evolved as a new form of communicative behaviour is given by inclusive fitness theory (an ultimate causal explanation). Yet an investigation of the transition from animal to linguistic behaviour requires above all a discussion of the possible proximate mechanisms that were involved. I shall first briefly recapitulate inclusive fitness theory and then discuss how linguistic behaviour evolved out of behaviours (also displayed by other animals) subsequent to bipedal loco-motion. My discussion of the proximate mechanisms will, however, be limited. I do not discuss the essential changes that led to the remarkable vocalizing powers that humans have, such as the changes in the position of the larynx. I also do not discuss in detail the transitions during infancy that result in the ability to speak and understand a language, i.e. from cooing (producing vowels) to babbling (involving vowels and consonants), gaze following, turn-taking and using words and sentences (see among others Fitch, 2010; Hurford, 2012; Locke, 2006; Tomasello, 2008).

If language was initially a new form of communicative behaviour, then language evolution is arguably explicable in terms of both *altruism* and *mutual benefit* (two principles of social evolution; see Hamilton, 1964; West, Griffin and Gardner, 2007), because linguistic behaviour probably enhanced inclusive fitness as the result of cooperation within the family and group. Cooperation may have *indirect* and/or *direct* benefits. It is probable that both have played a role in language evolution because language evolved in relatively small populations. Indirect benefits are explicable in terms of kin selection – limited dispersal, kin discrimination and the green beard effect are proximate mechanisms here. Direct benefits are explicable in terms of mutual benefit – cooperation as a by-product (of an otherwise self-interested act) or enforcement (through reward and punishment) are proximate mechanisms in the case of non-human animals. In the case of language evolution, it is thought that two other mechanisms are probably important, namely (direct and indirect) *reciprocity* (helping another because he or she will then help back, or because he or she has a good reputation) and enforcement through installing *social norms* (associated with punishment if someone violates the norm; see, e.g. Clutton-Brock, 2009). The effects of reciprocity and social norms are visible in current human social behaviours: humans cooperate within groups and the reciprocal exchange of resources between non-kin is widespread. These mechanisms are absent in other animals because the use of a *tensed language* is constitutive here (see Chapter 2, and Hacker, 2001). One reason why a tensed language is essential is that cooperation based on reciprocity often involves considerable time delays between assistance given and received. Because of these time delays, there are extensive opportunities for cheating. It is therefore thought that a tensed language evolved because it created new opportunities for cooperative behaviour (for it created opportunities to establish and pursue common goals over a period of time), but also 'solved' the problem of cheating through making agreements, installing norms, etc. These social arrangements are meant to discourage cheating and provide for sanctions against violations of rules. Other animals, by contrast, are restricted to cooperative strategies that generate benefits to inclusive fitness by pursuing goals in the immediate environment in space and time. It is, however, important to note that the evolution of reciprocity and social norms and the resulting differences between humans and other animals are the *outcome* of the long road to language evolution. Reciprocity and acting on norms are, just as intending, imagining, thinking, etc., examples of sophisticated linguistic behaviours that evolved during the later stages of language evolution. Focusing on these forms of cooperative behaviour does not help us to understand how the early communicative behaviours evolved. In the following, I shall investigate the first steps that led to language evolution.

Language is characterized by the oral or vocal flexibility: humans can produce speech consisting of different combinations of various phonemes. Yet language probably did not originate in oral flexibility, but in flexibility of gesture (see, among others, Tomasello, 2008). This hypothesis is based on the observation that vocal flexibility, in contrast to flexibility of gesture, is limited in chimpanzees (and was probably also limited in the early hominids). Ape vocalization is relatively stereotyped and linked to communicative signals and to particular expressions of emotions and needs. Gestures, by contrast, are relatively flexible in apes. For example, chimpanzees can use different gestures to achieve the same end, and they can use the same gesture in different contexts. This difference between flexibility of gestures and stereotypic behaviours is also visible in differences between chimpanzee groups: whereas some gestures (like the hand out begging gesture) are 'universal' among chimpanzees, other communicative gestures are variable from group to group. Notice that flexibility of gesture also explains why we can teach apes some gestures of a sign language (but not a spoken language). Oral flexibility (probably as the result of selection of variations of genes such as FOXP2, because the product of this transcription factor is, among other things, involved in the control of fine orofacial movements, enabling the hominids to develop articulate speech; see Enard, 2011) was therefore preceded by, or co-evolved with flexibility of gesture.

Bipedal locomotion extended flexibility of gesture because it freed the hands from the constraints imposed by quadrupedal locomotion: hands could be used in a wide range of new contexts unrelated to their prior functions. Accordingly, they could undergo new selection forces relating to gathering, extraction, sharing and processing of food; and to the fabrication, transportation, manipulation or exchange of various types of tools (Stanford, 2003; Chapais, 2008). Because flexibility of gesture was already present in the early hominids, bipedal locomotion enabled hominids to elaborate and refine behaviours that were occasionally displayed by their predecessors, and enhanced therefore the communicative skills of the hominids. An important gesture in the context of language evolution is pointing. This gesture enabled hominids to engage in triadic interactions and facilitated therefore the development of joint attention and action (i.e. pursuing common goals efficiently). One can easily imagine that the hominids began to use words because pointing at objects and activities could be used for explaining words. Hence the use of words probably evolved in a context in which pointing was already involved in communicative behaviours. For example, in the context of ordering ('take ☞') or demanding ('want ☞'), or for asking and expressing preferences: 'want (☞) this or (☞) that'? Hence the use of words probably evolved subsequent upon the use of gestures like pointing if the resulting new forms of communicative behaviours enhanced inclusive fitness. Notice that pointing has an important role in our current discourse: we use pointing for explaining word meaning through ostensive definition (e.g. 'this ☞ is red', while pointing at a tomato, 'this ☞ is walking', while pointing at an activity).

Yet although the use of pointing was probably an essential step from gestures to the use of words, this step was preceded by other psychological adaptations. These adaptations are important for understanding how pointing obtained a function for explaining symbols. They evolved when the hominids started to use tools and acquired the ability, through using their hands, to make and improve tools. The hominids began then to understand that a tool is good for achieving a specific goal. Chimpanzees choose suitable objects as tools from the immediate environment to achieve a certain goal. They sometimes modify the tool to improve its properties, enabling them to harness the concrete goal, but they tend to discard the tool after they have used it. The difference with the hominids is that after bipedal locomotion, hominids started to keep, transport and store tools and began to use them for permanent functions. Thus the permanent use of tools co-evolved with a shift in their use: they started to use the same tools in different contexts. Notice that the use of hands in the context of tool-making presupposes many adaptations, such as motor skills, hand–eye coordination, mnemonic skills, learning, ability to delay responses, etc. Making and improving tools requires several other adaptations, for example, that the hominids could attend to an object by inspecting it, so that they could learn more about its properties. When these adaptations evolved resulting in the use of self-made tools for foraging and hunting, one can imagine that their 'awareness' of the environment changed (see Holloway, 1981; see also Stout and Chaminade, 2012) as did the way they communicated about the environment. Csibra and Gergely (2006; 2009) argue that when children grew up in a group that used tools extensively, a system of *natural pedagogy* evolved enabling children to learn the function of tools through observing the behaviour of their parents. In contrast to their predecessors (and chimpanzees) children could then learn the function of tools because they could rely on the pre-existing knowledge about the tools of their parents. Evolutionary theorists have shown that social learning is in certain circumstances more profitable than relying on learning mechanisms like stimulus enhancement and emulation (see, e.g. Boyd, Richerson and Henrich, 2011).

The communicative behaviours involved in natural pedagogy were *ostension, referencing, gaze following* and *imitation*. These behaviours are independent of language and, hence, preceded language evolution. During child development, they are precursors of joint attention and action (and hence of social cognition). These communicative behaviours assist pedagogy if there is a *division of labour* between tutor and pupil, for teaching and learning are possible if both teacher and pupil understand who is emitting the information and who is supposed to pick the 'intended' information up. The child has to observe the (eyes of the) parent in order to extract the relevant information, whereas the parent has to observe the child in order to see whether the information is conveyed. One can imagine that subsequent to the use of the first words, talking to children (*motherese*) strengthened the conveying of information. It is believed that natural pedagogy resulted in social and cultural learning (i.e. the ability of children to adopt new forms of behaviour and utterances by watching and imitating others). Imitation also occurs, of course, in other animals, but imitation in animals is not an example of social learning: animals imitate behaviour because there is an immediate reward involved. For example, when the blue tit learnt through copying to peck at the foil tops of milk bottles, there was a reward: they obtained the cream on the top. And since pecking is something that these birds do anyway, it can be described as an example of using an existing behaviour in a slightly new context. Social and cultural learning based on imitating, by contrast, enables humans to copy the best among a number of alternatives and enables them to improve existing acts. Interestingly, the role of communicative behaviours involved in natural pedagogy may explain phenomena which are not well understood by developmental psychologists, such as 'persistent imitation' (Baldwin, 1892), 'rational imitation' (Gergely, Bekkering and Király, 2002) and 'over-imitation' (Lyons, Young and Keil, 2007). Natural pedagogy may also explain the occurrence of the perseverative search error (also called the A-not-B error), i.e. the observation that children of about 12 months of age tend to look for an object under container A (where the object was being hidden), despite the fact that they have seen that the object was replaced and hidden under container B. Studies have shown that this 'error' depends on the communicative interaction (eye contact, ostensive referential signals) between the child and the person replacing the object. If the test is done without communicative interactions, children make significantly fewer mistakes (see Topál *et al.*, 2008), showing that the eye contact and referential signals provided by the person are used by children as relevant information. One can imagine that they use the provided information for inferring that 'this kind of object is stored in container A', or that 'objects are kept in A', and this explains why they make the A-not-B 'error'.

The absence of natural pedagogy in animals also clarifies why studies of behavioural traditions in wild chimpanzees have not revealed evidence that these behavioural traditions accumulate improvements over time (see among others Tomasello and Call, 1997; Tennie, Call and Tomasello, 2009). Chimpanzees appear to focus on different information when they imitate gestures. For example, when chimpanzees watch someone using a tool, they tend to focus on the effect being produced in the environment, but pay little attention to the actual actions of the tool user. Chimpanzees use the observations mainly for producing the environmental effect themselves (e.g. using a stick for fishing ants).

The evolution of psychological adaptations involved in tool-making, teaching and learning and understanding the function of tools, clarifies why the use of words evolved when hominids, in contrast to their predecessors, could engage in communication about properties of objects. Pointing was important here, but it also involved better perceptual abilities for selecting attributes of tools, increased length of concentration and span of attention, the ability to convey information about the properties of objects, etc. The fact that there were many adaptations involved explains why it took millennia before the first use of words evolved. Yet these adaptations were important for language evolution, because the shift in attention helped hominids to focus on and to communicate about specific properties of objects. They understood then that the use of ostensive gestures (and thus the object of the actor's attention) was aimed at explaining a property of an object or tool. And explaining a property comes close to explaining a word (e.g. that pointing at a tomato is part of the explanation of the word 'red'). Hence the shift in focus of attention was a prerequisite for understanding that the aim of the deictic gesture is to *explain* something (e.g. a property or word). This helped them to understand that the object pointed at may have a function or communicative role beyond the immediate interests of the actor and recipient. Chimpanzees can focus their attention on something when someone is pointing at it, but they are only interested in whether the pointing results in something benefiting their own interests (e.g. something edible). Children, by contrast, begin to understand during their first two years that pointing also involves a social component: they begin to understand that pointing can be used to inform others and understand that pointing is used to inform them.

From the first use of words to speaking and understanding a complex language with a grammar is a long road. This is unsurprising given that the use of an expanded language requires several skills. Just observe our current linguistic behaviour and notice and realize what language enables us to do with words: we can use a tensed language enabling us to think of specifically dated events, of the past and the future; we can seek and discover general laws of nature and can imagine how things might be and how they might have been; we can talk about things although they are not present in the here and now and can imagine and fantasize about possible events and a promising date; we can pursue goals with the help of symbols, and this enables us to develop purposes beyond the temporal and spatial present. I have only discussed the first steps leading to language evolution and it is a topic for future studies to disentangle the subsequent steps.

Investigating the sequence of the steps that resulted in simple and complex linguistic behaviour is also interesting for another reason. The scenario discussed above raises the problem if and how the transition from instinctive to linguistic behaviour can, in part, be understood as an example of the transition from 'costly signalling' to 'cheap talk'. A well-known example of a costly signal is the peacock's tail. It evolved in the (competitive) context of sexual selection because a costly signal is thought to be a reliable signal: only individuals of 'good quality' can afford to invest in the costly trait and that explains why the receiver responds to the signal. Language probably evolved because this new form of communicative behaviour enabled individuals to coordinate actions in cooperative contexts enhancing inclusive fitness. Language is in game-theoretical terms an example of 'cost-free signalling'. The costs are paid by those who *deviate* from the evolutionary stable equilibrium of the coordination game (see Lachmann, Számadó and Bergstrom, 2001; Scott-Phillips, 2008). Yet the system of cost-free communication requires that individuals are able to check whether 'signals' are reliable with relative ease. It also requires that individuals can impose costs on those who are not reliable, for example, through social exclusion. A tensed language can strengthen 'cost-free communication', for individuals would then be able to exchange information about whether others were, are or will be reliable partners. An interesting point to notice is that these game-theoretical models allow 'signals' to take an arbitrary form. The reason is that the requirement is removed that costs must be causally associated with the signal form. The form of the signals in coordination games depends solely on whatever the signaller wishes to use as a signal.

Skyrms (2004, chapter 5) and others have argued that modelling the transition from costly signalling to cheap talk with the stag–hare game may be helpful, because they assume that language probably evolved in the context of cooperative hunting. In this game individuals can choose to hunt a stag or a hare. The stag (providing more meat) must be cooperatively hunted, while the hare can be hunted individually. Models show that cooperation (both individuals choose to hunt the stag) can be a stable outcome if they are able to coordinate their actions, and this may be one reason why cheap talk evolved.

7.7 Earlier weaning and cooperative breeding

When hominids started to use tools (for digging tubers and hunting big game), essential changes occurred in human life history (i.e. the timing of key events in an organism's lifetime). Earlier weaning is a well-known example. I discuss possible effects of imprinted genes on earlier weaning and on the development of gestures and linguistic behaviour that co-evolved with earlier weaning.

When food (as the result of hunting and gathering) was present on a regular basis, hominids started to wean children at an earlier time. Humans have nowadays a much shorter interbirth interval than our closest relatives, but have a longer juvenile dependence (Bogin, 1997; Kennedy, 2005; Humphrey, 2010). Humans wean in natural fertility populations at about 1–4 years, while chimpanzees wean at 5 and orang-utans at 7.7 years. The transition to shorter interbirth intervals was made possible because humans started to use protein-rich meat (but also plant food and tubers) as a supplement to and alternative for maternal milk. Lactational anoestrus was therefore removed at an earlier age. Chimpanzees are not able do this because there is far less reliable food provisioning in chimpanzees. Chimpanzee infants derive their protection and nutriments from their mother, and weaning at an earlier time would probably reduce inclusive fitness since infants lack the skills to survive on their own in an environment in which food availability is fluctuating. Fruiting trees may for chimpanzees present times of super-abundance, but they are interspersed with long periods of want. In these periods of want chimpanzees rely on fishing insects as the main source of rich nutrition (fishing insects has to be learnt by infants).

Early weaning had an enormous impact on fitness: the number of children women could conceive increased. A simple calculation shows why human populations started to expand as the result of earlier weaning. Suppose that chimpanzee and human females can conceive children after they are 20 until they are 40 years old. Assuming lactational anoestrus and that chimpanzees wean at 5 and humans wean at 2.5 years, chimpanzee females can conceive four children while human females can conceive in the same period eight children. If 50% die in infancy, only the human population grows. Yet installing earlier weaning requires morphological and physiological adaptations of the child. For example, 'tiny incisors' (milk teeth) evolved enabling children to consume solid food at an early age (milk teeth are absent in other apes). It also required behavioural adaptations enabling mothers and children to adjust their communicative behaviour to early weaning. Since mothers had to feed and protect several children at the same time, Hrdy (1999; 2009) and others have argued that a system of cooperative breeding evolved (allomothers helping a mother raise her children). It is believed that new forms of non-verbal and later linguistic

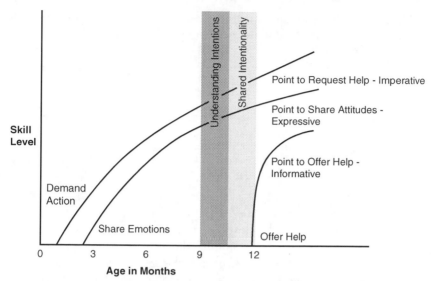

Figure 7.2: The development of skills necessary for collaborative action facilitated by the emergence of cooperative communication in pointing (reproduced with permission from Tomasello, 2008, p. 144).
It is better (or less misleading) to talk about 'understanding goals' and 'pursuing common goals' instead of 'understanding intentionality' and 'shared intentionality', for the formation of intentions is an example of a complex skill and develops later in children, i.e. during their fourth year.

behaviour evolved (some investigators hypothesize that babbling was one of these behaviours) because these behaviours enabled children to participate in the system of cooperative breeding (Burkart, Hrdy and van Schaik, 2009). The development of the cooperative breeding system was probably facilitated by natural pedagogy that evolved earlier (or co-evolved with cooperative breeding). For example, children could use communicative gestures involved in natural pedagogy like pointing to indicate their desires, and later use these gestures in the context of helping others (see Figure 7.2).

Inclusive fitness theory predicts that natural selection favours communicative behaviours (like pointing, babbling) if these behaviours optimize the functioning of food provisioning in the cooperative breeding system, but also predicts intragenomic conflicts. For if food is in short supply, children of different ages have to compete for parental investments. If a shorter period of breastfeeding is beneficial for matrigenes but disadvantageous for patrigenes, kinship theory predicts that matrigenes promote an earlier

Table 7.1: Possible effects of patrigenes and matrigenes (described in catchwords) on child development during the transition from breastfeeding to solid food.

Patrigenes	Matrigenes
Prolonged breastfeeding.	Earlier transition to solid food.
Instinctive sucking, crying and reactive crying.	Synaptic plasticity necessary for learning skills involved in cooperative breeding.
Attachment (happy disposition).	Detachment (exploration, vigilance).

transition from breastfeeding to consuming solid food, whereas patrigenes favour a longer period of breastfeeding (Smit, 2009; 2010d; see Table 7.1).

There is clear evidence that imprinted genes are involved in postnatal development. For example, the product of the patrigene MAGEL2 is a trigger for the onset of breastfeeding (Schaller et al., 2010). There is also evidence that the patrigene NDN affects the differentiation of mesenchymal (stem) cells in either myoblasts (muscle cells) or adipocytes (fat cells), for NDN appears to inhibit the differentiation of stem cells into (pre-) adipocytes (Bush and Wevrick, 2012; see also Tseng et al., 2005; Yu et al., 2000). Patrigenes may promote the development of a larger, muscular body so that children are better competitors, whereas matrigenes may favour a less muscular body which uses resources for increased deposition of fat, for children then have reserves and need fewer feeds during periods of food shortage (Haig and Wharton, 2003; Smit, 2006). This explains why children with Prader–Willi syndrome (PWS), lacking the expression of NDN, develop a less muscular body. An interesting gene in the context of this chapter is the matrigene UBE3A. The product of this gene affects synaptic plasticity in structures such as the hippocampus and cortex (Greer et al., 2010; Yashiro et al., 2009; see also Kühnle et al., 2013). Mice lacking the expression of the orthologous gene do not explore their environment. Children with Angelman syndrome (AS) (lacking the expression of UBE3A) do not explore their environment, do not develop the ability to engage in triadic interactions with parents, do not babble and hardly imitate non-verbal or verbal behaviour and do not develop speech (see Didden et al., 2004; Joleff and Ryan, 1993; see also Plagge et al., 2005 on the link between imprinted genes and exploration). Hence it appears that UBE3A contributes to the development of communicative behaviours enabling children to participate in the systems of natural pedagogy and cooperative breeding. In opposition to those who argue that children with AS have a deficit in 'reading the mind' leading to 'mind-blindness' (see Baron-Cohen, 1995; Badcock and Crespi, 2006; 2008), I have argued that they have a deficit in the ability to acquire the early communicative behaviours (including the first forms of linguistic behaviour) which evolved because of earlier weaning (Smit, 2009; 2010a; 2010d). This explains why mutations of the matrigene UBE3A have devastating effects, for the gene is now constitutive to what we now describe as the innate ability of children to acquire a language. As I have explained above, language evolution required several psychological adaptations (changes in gesturing, attention, learning, imitation, vocalizing, etc.) and these culminated in the second-order ability to acquire a language.

UBE3A was selected since its effects enhanced the inclusive fitness of matrigenes as the result of earlier weaning and because the effects of the gene enabled children to participate in the system of cooperative breeding. Yet it is possible that the (earlier) development of communicative behaviours was favoured by matrigenes for another reason: the system of natural pedagogy enabled the mother to educate and mould the child's behaviour according to the mother's interests. Recall that the early communicative behaviours that evolved were asking, demanding, ordering, etc. Because fathers are less involved in education during the early stages, patrigenes are thought to rely on affecting instinctive actions and reactions like crying and reactive crying (see Smit, 2009, 2010d).

7.8   Conclusion

Humans can express their thoughts in linguistic behaviour. They can think of dated events, of the past and the future, of what exists but also of what does not. They can also think of how things might be and how things might have been, and have therefore the power of creative imagination. They are self-conscious creatures since humans are able to reflect on their motives, affections and thoughts, and can report about their reflections.

I have argued that these typical human abilities are examples of complex skills that evolved subsequent upon the evolution of more simple linguistic skills. Asking how these 'higher-level' abilities evolved (e.g. how consciousness evolved, or how language enabled humans to translate thought into the medium of language for purposes of communication) ignores these simple skills and leads to attempts to explain how language, as it were, directly evolved out of brain processes (how does the brain produce consciousness or knowledge of language?). It results in an incoherent conceptual framework for understanding the evolution of what Chomsky called the language faculty. Instead of investigating how the mind/brain possesses knowledge of grammar and how this knowledge is generated by the brain, it is far more interesting to start with the question of how the ability to do things with words evolved. For the primary function of language during our evolutionary history was not the communication of thought, but to engage in communicative behaviour. Telling another what one thinks is an advanced part of human communication that evolved only subsequent upon the evolution of the ability to engage in activities like asking, demanding, ordering, expressing emotions and so forth. And the ability to do things with words was preceded by, or co-evolved with, flexibility of gestures and the goal-directed use of the hands. Because flexibility of gesture was present in the early hominids whereas vocal flexibility was absent, linguistic behaviour evolved as the continuation of flexibility of gesture. I have argued that bipedal locomotion can be seen as one of the essential steps on the road to language evolution, because it freed the hands from the constraints imposed by quadrupedal locomotion. Tool-making, the permanent use of tools, pointing as a communicative gesture, natural pedagogy and several psychological adaptations, led to the use of words when the hominids understood that pointing could be used for explaining something beyond the immediate interest of the actor and recipient. The starting point was flexibility of gestures; the result was vocal flexibility. I have also argued that language evolution was made possible by several genetic adaptations and expect that future studies will provide us with more insight into the effects of genes that shaped language evolution.

Investigating how linguistic behaviour evolved out of animal behaviour helps us to understand why humans became unique creatures among the apes. Humans are special because we have an innate ability to learn a language and this enables us to develop a distinctive array of rational powers of the intellect and will (see also Hacker, 2007). Of course, other animals are smart too. Chimpanzees are able to learn from others how to crack nuts and how to use sticks to fish for termites; Japanese macaques learn how to wash potatoes. Yet there is accumulating evidence that there is a difference here between humans and other animals: only humans possess the ability to adopt new forms of behaviour and utterances by watching others. This ability is rooted in social or cultural learning and resulted in linguistic behaviour and later in the accumulation and improvement of symbolic knowledge. This chapter shows why cultural evolution required many genetic adaptations during the early stages of language evolution. These genetic adaptations paved the road to the evolution of a unique culture-creating species.

CHAPTER 8

Moral behaviour: a conceptual elaboration of Darwin's ideas

8.1 Introduction

In *The descent of man* Darwin argued that:

> a moral being is one who is capable of comparing his past and future actions or motives, and of approving or disapproving of them. We have no reason to suppose that any of the lower animals have this capacity; therefore when a monkey faces danger to rescue its comrade, or takes charge of an orphan-monkey, we do not call its conduct moral. But in the case of man, who alone can with certainty be ranked as a moral being, actions of a certain class are called moral, whether performed deliberately after a struggle with opposing motives, or from the effects of slowly-gained habit, or impulsively through instinct. (2003 [1871], pp. 111–112)

In his explanation of why only humans evolved into moral beings Darwin used both evolutionary and conceptual arguments. Darwin argued that moral behaviour is to a large extent social behaviour and that its evolution in humans and other animals can be understood in terms of his theory. He used individual and group selection in order to understand the roots of social behaviour. Yet Darwin also argued that there is an essential difference between humans and other animals: according to him, other animals lack the mental capacity to develop moral behaviour because they cannot, in contrast to humans, refer to past events or future plans as motives for their actions. Accordingly they do not have or experience moral emotions such as remorse and indignation.

In this chapter, I shall elaborate Darwin's observations of moral behaviour. I shall argue that we can understand why only humans evolved into moral creatures if we recognize the constitutive role of language in the evolution of moral behaviour. I shall confront the model I advance with two other attempts to explain the role of cognition and emotions in moral behaviour: emotionism and nativism. According to the emotionist models, emotions affect our moral cognition; according to the nativist models, emotions follow moral cognitions. I shall argue that these models are conceptually incoherent and lead therefore to misinterpretation of empirical data.

I start with a brief discussion of Darwin's ideas and shall explain how we can elaborate these ideas through invoking language evolution. Next, I shall use the elaborations of Darwin's ideas for understanding the evolution of inbreeding avoidance and the incest taboo. What are the differences here between emotionism, nativism and the model I advance? Finally, I shall discuss in detail why the emotionist and nativist model are conceptually incoherent, and use the evolution of fair sharing in order to show how the third model can be used for developing testable hypotheses.

8.2   *Homo loquens* and moral behaviour

Darwin argued that only humans are moral beings. In his explanation he discussed the example of a swallow experiencing a conflict between two instincts: the instinct to care for offspring and the instinct to migrate. Suppose that the migratory instinct is 'more persistent' and gains the victory. The swallow then takes flight and deserts the offspring, presumably when they are for a brief moment out of sight. But when the bird arrives at the end of the journey and the migratory instinct ceases to act, she will not feel the agony of remorse, according to Darwin, and there is no 'image constantly passing through her mind, of her young ones perishing in the bleak north from cold and hunger' (Darwin, 2003 [1871], p. 113). Humans may also follow, according to Darwin, the stronger impulse, and this will sometimes lead humans to gratify their own desires at the expense of others. But since humans have the ability to judge their actions, they may later realize that they should have behaved less selfishly. They may feel remorse, repentance, regret and shame (possibly related to the judgments of others). Humans can then resolve to act differently in the future, for only humans have the mental capacity to look backward and forward. And because humans understand the difference between right and wrong and are capable of wishing that they had not done what they did (hence they understand counterfactuality; see von Wright, 1974, pp. 39ff.), they can use their conscience as a guide for the future.

Darwin's observations raise the question of how we understand why only humans have the capacity to look backward and forward and are able to judge their actions. The answer, I suggest, is that only humans are language-using creatures. The behaviour of other animals is, of course, not only determined by their instincts, for animals can learn and know therefore many things. But the cognitive powers of animals are limited since they do not use a tensed language (see Chapter 2). For example, animals have wants and can pursue the objects of their desires (water, food, sex), but the horizon of their wants is limited since the objects of their will are constrained by their limited mental capacities: they can choose and have preferences, but cannot deliberate. For animals do not use a language, cannot give reasons for their preferences and cannot reflect on these reasons. Hence when we say that animals make a decision, we mean that they simply terminate a state of indecision (see Darwin's discussion of the swallow), since they cannot explain and justify their decisions (and actions consequent upon decisions) by forward- and backward-looking reasons. By contrast, humans can express their reasons and can do something because it is desirable or obligatory given certain reasons or values. Suppose that my mother orders me to bring a pie to my uncle tomorrow at 10.00 a.m., then I am able to form the intention to do what she asked me to do. The formation of this intention is not possible in other animals because they do not use kin categories and a tensed language. Hence animals cannot explain their goals by referring to kin categories and backward- or forward-looking reasons, and cannot explain why it is desirable or obligatory to bring food to a relative ('I promised my mother to bring a pie to my uncle') because they cannot justify their behaviours by reference to reasons.

The invention of language differentiated humans from other animals. It enabled humans to develop new forms of communication and has freed humans from their primate heritage: natural selection was extended with cultural evolution. Yet it is probable that natural selection has shaped or canalized the transition from nature to culture. Suppose that the early hominids invented language and started to use kin categories (mother, child, etc.). The use of these categories affected then the way they interacted with kin and non-kin. Kinship theory predicts that natural selection will shape linguistic behaviour if the use of kin categories enhances or reduces inclusive fitness (assuming that there are genes affecting behaviours in which linguistic categories are invoked). Hamilton (1975) thought for this reason that our genetic system does not provide a blank sheet for cultural development. The genetic system generates, according to him, various inbuilt safeguards and the sheet is lightly scrawled with certain tentative outlines. An example of a *tentative outline* is that linguistic categories were invented by a species that was already able to engage in cooperative behaviour with kin. Mothers learn to identify their children after birth and care for them; children learn to recognize their mother (through what Konrad Lorenz called imprinting) and demand care. Hence when the early hominids started to use kin categories and names in a context in which kin selection had already shaped behavioural propensities, they could use these to enrich their behavioural repertoire, for example, for developing shared plans and projects. For example, Hrdy (1999; 2009) has argued that the development of cooperative breeding in humans (for example, grandmothers and other mothers helping a mother raising her children) evolved as an extension of existing forms of parental care and required new forms of (linguistic) communication. An example of a *safeguard* may be the following. Suppose that the early hominids invented kin categories and names for individuals. Because of their perceptual and mnemonic abilities, they could use these words for communication. The use of pointing may have been a safeguard here. For example, when a mother asked another female in her extended family or group to look after her eldest daughter for an hour, she could also point at her to focus the attention on whom she meant (reducing the risk of mistakes). Other animals, moving in a quadrupedal manner, do not use pointing as a gesture involved in joint attention and triadic interactions, and do not develop shared intentionality (Burkhart *et al.*, 2009; Tomasello *et al.*, 2005).

The result of the invention of language was that humans became rule-following creatures. For only language-using creatures can follow a rule: they use rules as *reasons* for what they are doing and are therefore capable of recognizing and explaining that they have made mistakes through referring to the rules. It is important to notice that the ability to use rules is an example of a complex skill. It takes about three years for children to acquire this skill, and it took the hominids probably several millennia to develop a rule-governed use of symbols. This explains why the evolution of language gradually resulted in a gap between animal and human behaviour, between nature and culture. It also explains why humans started to act on norms after they developed the ability to use rules and, hence, after they developed complex linguistic skills. A brief discussion of how children acquire moral behaviour as an extension of intentional behaviour illustrates this.

Children become moral creatures during the first four years of their life (see Hacker, 2007, chapter 8; Smit, 2010b; Chapter 5 of this book). It starts when children, exploring their environment, begin to answer the what and why questions posed by their parents. ('Why are you doing that?'; 'What do you want to achieve?') By answering these questions, children learn to describe the purpose of their behaviour and to justify it. Hence they must already be able to speak and understand words and sentences. Next, they learn to announce an action ('I am going to V') and learn then that they must go on to V, for that is what they said they were going to do. If a child declares that he is about to throw a ball to another but never does so, then he does not yet understand what intentional behaviour is. Later on, children learn to give reasons for what they want to do, i.e. they form primitive intentions. Hence the formation of intentions goes hand in hand with learning to give reasons, reasoning and learning what count as adequate reasons. When they, during their fourth year, learn to specify why certain actions are desirable (or permissible or obligatory), they acquire a moral sense. They become then moral creatures who can be held responsible for their deeds.

The observation that the development of moral behaviour takes time during ontogenesis and took time during our evolutionary history distinguishes the model I advance from the other two models. Emotionism states that the affective system, shaped by natural selection, floods our consciousness with emotions that intuit the standards of good and evil. Emotions are for this reason thought to *cause* our moral judgments (hence the development of linguistic abilities is thought to be of minor importance). Nativism assumes that, as the result of a mutation, the human brain was rewired such that the moral faculty was acquired *suddenly*, resulting in an innate knowledge of right and wrong. I shall elaborate the differences between the three models through discussing how they explain the evolution of the incest taboo.

8.3 The evolution of the incest taboo

The *emotionist* interpretation of the evolution of the incest taboo is an elaboration of a hypothesis advanced by Westermarck (1891, and later editions; see also the discussion in Wolf and Durham, 2005). Westermarck argued that natural selection has resulted in a natural tendency to avoid the damaging effects of inbreeding. He at first thought that there was an 'instinct' causing inbreeding avoidance, but later modified his view (for incest avoidance is, at least in part, acquired) and argued that it was not an instinct but an 'acquired aversion' (e.g. Westermarck, 1925 [1891], p. 197). Westermarck thought that close kin experience a *natural aversion* to engage in a sexual relation as the result of developmental familiarity. According to him, experiencing this emotion led to moral disapproval and prohibitory customs and laws (see also Westermarck, 1906, chapters 1–5). Later theorists (see, for example, Arnhart, 2005) argued that the incest taboo evolved as a *generalization* of this emotion. Hence the problem arises how we can explain the evolution of this generalization, i.e. the alleged transition from an experienced emotion to a moral rule. Chapais (2008, pp. 83–86) discussed the following scenario. Suppose that an early hominid witnessed a sexual act between two opposite-sex siblings, and assume that he was able to transpose this observation to his own situation, i.e. he also formed a mental image of a sexual interaction with his own opposite-sex sibling. Because he experienced an aversion in his own situation (as suggested by Westermarck), it was, according to Chapais, possible that he also developed an aversion concerning the sexual act he witnessed. This would, according to Chapais, be an example of empathy and depended on the possibility that hominids could recognize similarity between the two sexual bonds. If all individuals in the tribe had this similarity experience, the aversion would be shared collectively. And if these individuals could use a primitive language, they could communicate about the shared aversion. The shared aversion would then lead to the differential treatment of the 'wrongdoers' because of the shared aversion (avoiding or rejecting wrongdoers). This behaviour towards wrongdoers is, according to Chapais, close to acting on a norm.

However, the emotionist explanation faces several problems. First, the empirical problem that there is no clear evidence that incest avoidance is caused by a natural aversion. Westermarck used the term 'aversion', but what the data show is that developmental familiarity increases the probability that individuals are 'less attracted' to each other or 'indifferent' with regard to a sexual relation (Shor and Simchai, 2009; see also Leavitt, 1990; 2007). This observation raises the question why Westermarck used the term 'aversion' (and why others use the word 'disgust'). Since we express moral emotions when we are confronted with concrete violations of moral rules (see further below), I suggest that Westermarck used this observation as an argument for the hypothesis that emotions were selected as an avoidance mechanism, i.e. he projected the moral emotion into the course of evolution, resulting in the misguided idea that the incest taboo evolved as a generalization of an alleged natural aversion. Second, there are evolutionary arguments against the idea that natural selection always leads to an aversion. Haig (1999) has argued that incest can increase an individual's inclusive fitness. He showed that, assuming that the relative fitness of an inbred child is 0.58–0.80, incest enhances the inclusive fitness in certain circumstances. Hence Haig's model does not predict a 'natural aversion' in all circumstances. The model also predicts differences between the sexes: since a sexual relation has higher opportunity costs for females than for males (see Trivers, 1972), females are predicted to be more 'averse' to incest than males. Third, in Chapais' scenario a hominid is capable of imaging himself copulating with an opposite-sex sibling, but the problem with this scenario is that it presupposes the ability to use a tensed language. For it is the intellect that confers meaning on the image, not the visual image on the thought (as the empiricists mistakenly believed; see Kenny, 1989, chapter 8). If a hominid could not use a tensed language to express and report about his images, then it is inconceivable how he knew whether his image of himself copulating with an opposite-sex sibling was of what he had done, what he wanted to do, what he was going to do, what he should not do, and so forth. Fourth, it is unclear how emotional feelings inform us of good and evil. How does a physiological response of the body cause knowledge of right and wrong?

The *nativist* interpretation of the evolution of the incest taboo assumes that incest avoidance is brought about by an innate linguistic system in the brain. This system is thought to consist of innate kin categories linked to a tendency to avoid incest. These linguistic kin categories are said to be extensions of 'natural categories' already present in the mind/brain of our predecessors and other animals. For example, Fox (1975; 1980, chapter 7; 1989, chapter 9) argued that the discriminatory abilities of our predecessors and other animals created 'category systems' which, while more limited than linguistic systems, operated in much the same way as the system based on the use of language. He used Chomsky's principle and parameter model for explaining how 'linguistic categories' evolved out of these 'natural categories'. Recall that Chomsky argued that children possess an innate language faculty that consists of an abstract Universal Grammar incorporating a set of principles. Following Chomsky, Fox argued that humans possess innate categories in their mind or brain for distinguishing kin. These categories are associated with tendencies to treat individuals differently: 'If these tendencies that we have described are truly universal, like the categories of the universal grammar – if … they are species-specific tendencies – then they exist in the human mind for the same reason the bones exists in the human body and the cells in the brain' (Fox, 1980, p. 183). However, the nativist model is confronted with three interrelated problems. First, Fox thought that humans possess natural categories in their brain, although these do not 'determine' the use of words. For these categories belong, in terms of Chomsky's model, to the principles. This idea is highly problematic for conceptual reasons, for there are no such things as 'natural categories' in the brain or mind (see Hacker, 1990a; Chapter 6 of this book). The fact that animals respond differently to 'mothers' and 'sibs' is not an argument for saying that there are 'natural categories in the brain'. Second, Fox argued that the use of 'linguistic labels' evolved as an extension of 'natural categories' already present in the mind/brain of the predecessors of humans. He thought, just as Chomsky, that the innate categories 'grow' in the mind/brain and 'give voice' via systems of linguistic classification. This is problematic, for it is unclear how 'natural categories' can 'grow and give voice' in the brain. Some things can grow in the brain (e.g. a tumour), but if someone's knowledge grows, then this knowledge does not grow in someone's brain. It is better to argue that, during child development, natural behaviours are replaced by linguistic ones. For example, children learn to replace the natural expression of fear (crying) with linguistic expressions, for example, 'Mama!!!' or, 'I'm frightened.' This extension enables children, in contrast to other animals, to answer what they are afraid of (e.g. 'A scary dog'). This ability to specify the reasons for their emotion (the child can explain that a dog bites and, hence, why it wants to run away) is not present in animals and creates new possibilities for acting and for communication. Third, because 'linguistic labels' and moral rules are seen as extensions of natural categories and tendencies, Fox thought that the development of language was of minor importance for understanding the evolution of the incest taboo, for installing this taboo represented nothing more than a 'labelling' procedure for behaviour that would occur anyway. This is highly problematic, because Fox conflates here avoiding incest caused by a 'natural tendency' and avoiding incest because an individual acts on the norm, 'Incest is wrong.'

8.4   A taboo superimposed on a disposition

According to the alternative model advanced here, the incest taboo evolved as a norm superimposed on a natural disposition (inbreeding avoidance as the result of developmental familiarity). The incest taboo was probably installed in order to avoid the damaging effects of inbreeding. I assume that humans started to use the norm after they had developed sufficient linguistic skills to employ the moral rule, 'Incest is wrong.' For humans must be able to recognize the truth of this rule and that requires several skills (e.g. being able to display intentional behaviour, understanding the effects of acts and that acts may be beneficial or detrimental for the health and well-being of themselves or others, being able to differentiate between right and wrong, good and evil). Durham (2005) has investigated this possibility and found that in 54.8% of the studied societies people mention some form of recognition of inbreeding effects as a reason for the incest taboo. In the remaining 45.2% there was no information available (hence it is possible that people also recognized these effects as a reason in these cases). Durham insists that the harmful effects (on offspring) may have been one of the reasons to install a moral rule, for incest also has harmful psychological effects (on parents, victims) and is therefore disruptive for relations within the family and group (hence these effects may also have been a reason to install the rule). I discuss two elucidations.

First, note that this explanation is at variance with the hypothesis advanced by Freud (1950 [1913]) and others, namely that the incest taboo was installed to suppress sexual relations in the family. If the model I advance is evaluated in terms of the Freud–Westermarck debate (discussed in Wolf and Durham, 2005), then I take sides with Westermarck. I also assume, in contrast to Freud, that there are during the early years (the first 5–7 years) no sexual bonds between, for example, brothers and sisters but asexual bonds. Just as Westermarck, I assume that these early asexual bonds affect the probability negatively that individuals engage in sexual relations later in life, for there is empirical evidence showing that children, raised together, avoid a sexual relation later in life even if they are not kin (see, among others, Lieberman, Tooby and Cosmides, 2007; Wolf and Durham, 2005). I ignore here the fact that there may be differences between the sexes (and intragenomic conflicts; see Haig, 1999). In contrast to Freud, Lévi-Strauss (1969 [1949]) and others, I also assume that the incest taboo was not installed as a prerequisite for exogamy. Freud thought that, because the early hominids were naturally inclined to mate and marry within the family, competition between males in the family (the sons competing with their father for sexual access to their mother) would lead to unstable social relations. In order to avoid struggle in the family, Freud thought that the incest taboo was installed in order to maintain stability and order. Lévi-Strauss later argued that the incest taboo led to exogamy, i.e. the cultural practice of exchanging mates between groups (resulting in alliances). The problem here is that female chimpanzees, for example, also leave their natal group (while there is no taboo in chimpanzees) and mate in another group. Hence it is better to argue that exogamy evolved as an *extension* of female and/or male dispersal already present in animals (see Chapais, 2008).

Second, note that I assume that developmental familiarity leads to an acquired *disposition* to avoid incest and argue that the incest taboo was superimposed on this disposition. I have two reasons for talking about a disposition (see Ryle, 1980 [1949], chapter 5). First, I assume that our predecessors, before language had evolved, acquired the disposition to avoid sexual relations with those with whom they were intimately raised (a dyadic phenomenon). Because childhood associates are in most cases close kin, developmental familiarity was for natural selection a 'mechanism' to decrease the fitness-reducing effects of inbreeding (but it is possible that there are other mechanisms involved; see Lieberman, Tooby and Cosmides, 2007). Second, I use the term 'disposition' to emphasize that avoiding incest is not a response to a certain situation, but that someone is bound to avoid incest because he has acquired the disposition. If he has acquired this disposition, then we expect him or her to display one sort of behaviour: he is not interested in engaging in a sexual relation with close kin. I assume that the incest taboo (a universal standard for moral conduct) was later superimposed on this natural disposition and evolved when humans were able to recognize and understand the damaging effects of incest. They developed then the moral attitude to care about the good (health and well-being) of their own and others. However, who count as close kin may vary from society to society and from one historical period to another, and this explains why the incest taboo has various applications.

When humans developed sufficient linguistic skills and started to use the rule, 'Incest is wrong', as a standard for moral conduct, they were able to *act on* a moral belief (further explained in Johnston, 1999). The rule told them how they should behave *and* how other humans should behave. Yet although they developed then a long-standing moral attitude, it is possible that they experienced conflicts between desires and how they ought to behave. Suppose that a father recognizes the truth of the rule, 'Incest is wrong', but is attracted to his daughter. This is not a fictive example: in one study 11% of non-incestuous fathers reported some degree of sexual arousal in response to their daughters (Williams and Finkelhor, 1995). Yet if the father holds this moral belief and exercises self-control, then he will see it as his moral duty to avoid a sexual relation. He feels perhaps horrified at his own desire (just as he would feel disgust when he hears about others having a sexual relation with their relatives) and these feelings tell him that he cares about this rule for moral conduct. But it is possible that he also experiences the temptation to do something that is prohibited by the rule. But if he is seduced by the temptation, then he will, as a moral creature, later realize that he should have behaved differently. I have chosen the example of a father, but empirical investigations have shown that there are differences here between the sexes: daughters are less attracted to a father than a father to his daughter. Most incestuous acts are unwanted by females and initiated by males.

If the incest taboo was superimposed on a natural disposition, then there are two causes or reasons for why incest occasionally occurs. First, the third model assumes that individuals, raised together, avoid a sexual relation later in life. Hence the model predicts an increased chance of incestuous relations when father and daughter, mother and son, brother and sister, have not been intimately associated in the early years. There is evidence in favour of this prediction (see, among others, Erickson, 2005). Second, the model predicts that humans do not engage in incestuous relations because they act on a moral belief. However, moral non-education does occur and humans can become uncaring, cynical, callous and even cruel. Notice that, if someone has learnt the difference between right and wrong and later becomes uncaring, that we do not say then that he or she *forgets* this difference (Ryle, 2009 [1971], Essay 29). The reason is that when someone learns the difference between right and wrong and acquires a sense of obligation, he or she learns to care about what is right and wrong and internalizes standards for moral behaviour, although these standards of behaviour may be lost (when someone becomes uncaring), explaining why incest occasionally occurs.

8.5   The emotionist model

Research into the ontogenesis and evolution of moral behaviour requires a correct conceptual framework characterizing the concepts of the phenomena investigated. In the following, I shall argue that emotionism and nativism do not serve this purpose because these models are confronted with conceptual problems. I shall also elaborate the alternative model further through discussing the example of collaborative action in detail in order to demonstrate how the model can be used for understanding the evolution of moral behaviour.

The emotionist models can be traced back to the ideas of James and Damasio. James (1950 [1890], vol. 2, chapter 25) argued that the feeling (of an emotion) arises as the result of bodily changes (e.g. increased heart rate). For example, suddenly encountering a burglar in the house causes bodily changes and these changes result, according to James, in an emotion: feeling frightened. Damasio (1996) later elaborated these ideas. He argued that the feeling of an emotion is brought about by a connection between an image or representation of a situation evoking the emotion and bodily changes. Emotional feelings provide us a glimpse of what goes on in the body when an image of these bodily changes is juxtaposed with the image or representation (of the situation evoking the emotion). By this juxtaposition, the body images give the representation a quality of goodness or badness. They become, according to Damasio, somatic markers enabling us to evaluate a situation. Hence the somatic markers, which our brains create as the result of conditioning at a neural level, are the 'gut reactions' which we experience when we are confronted with moral problems. They guide us when we encounter similar situations or problems. While James thought that the perception of the bodily changes caused the quick emotional response, Damasio believes that the acquired somatic markers become the guides for making quick judgments and decisions, and tell someone how he or she should evaluate the situation ('good or bad'). However, the model of Damasio is problematic for two reasons. First, the model mistakenly assumes that feelings of emotion are the result of juxtapositions in the brain of images of situations evoking emotions and (images of) bodily changes. A blush of embarrassment brings home to me that I am ashamed of having lied, but I do not *perceive* then images in my brain. Second, the model does not differentiate the cause of an emotion and its object (see Hacker, 2004; Kenny, 1963; Smit, 2010a). Suppose that I am lying in bed and hear a noise in the night. Then I am frightened because the noise may have been produced by a burglar. Hence, there is a difference between what I am frightened by (the noise) and what I am frightened of (the object of the emotion, i.e. being harmed by a burglar). This explains why the emotion of fear disappears if I subsequently notice that the noise in the night is produced by a cat, for the fear is then no longer warranted. Human emotions can therefore be reasonable or unreasonable because they involve an element of judgment and appraisal related to the object of an emotion. This ability to specify the appropriate object of an emotion differentiates children from other animals. Of course, emotions in animals have an object too: a monkey screaming in fear of a snake knows snakes to be harmful. But the cognitive and appraisive aspects of the objects of emotions are far wider in a language-using creature. Only language-using creatures can feel remorse or guilt for having done something that they knew to be wrong or can feel hope that another person will suffer less misfortune (of poverty or illness).

The models of Haidt (2001; 2007), Greene *et al.* (2001), Greene (2002) and Prinz (2007) are extensions of the ideas of James and Damasio. They assume that emotion and cognition are separate and partially independent systems in the brain. The cognitive system in the brain is said to be slow compared to the affective system (see Zajonc, 1984). Haidt uses these systems to explain the sequence in moral behaviour. Suppose that someone is confronted with a moral problem. Then there is a sudden appearance in consciousness of a moral intuition including an 'affective valence' ('good or bad', 'like or dislike'). This intuition is, according to the model, caused by the fast affect system in the brain and tells the person whether the issue at stake is right or wrong. The possession of the slow cognitive system explains, according to the model, why humans start with (what Haidt calls: post hoc) moral reasoning after the intuitions enter consciousness. Since this reasoning is often 'an attempt to justify the intuition', Haidt concludes that moral judgments are primarily caused by the affective system in the brain. This hypothesis is, however, problematic. Suppose that I am told that innocent people have been killed and experience a gut feeling. Then it is not the gut feeling that tells me that it is an unjust action, as Haidt assumes; it is the act I am informed about that is unjust because someone's rights are at stake. And since I care about the protection of rights of human beings, I feel indignant and this may be accompanied by a flush of indignation or a gut feeling. The bodily perturbations accompanying moral emotions show then that I am *not indifferent* with regard to certain goals, and inform me about my *attitude* towards a certain moral issue, just as a pang of jealousy may indicate that I am falling in love with a person and a blush may inform me that I am ashamed of what I have done. Thus my feelings are indeed guides to action related to good and evil. But not because the feelings cause or precede my judgments: my feelings are not the markers of good and evil or the causes of my moral intuitions. When children acquire the ability to pursue goals, then they acquire the capacity to care about right and wrong and their emotions help them then to pursue these goals efficiently. For one feels no emotions about things concerning which one is indifferent, and one does not pursue goals efficiently unless one cares about achieving them (see Hacker, 2007). This also explains why Damasio found that patients with damage to the ventromedial prefrontal cortex showed emotional deficits (while they retained their cognitive abilities, including IQ and knowledge of right and wrong) which crippled their ability to develop judgments and their decision making. Damasio argued that these patients lack the somatic markers as guides for right and wrong. Yet this interpretation is problematic, since emotions do not inform us about right and wrong. A better interpretation is that the brain damage affects the ability to care about goals and objectives. For it appears that the brain damage affects both the patient's emotions *and* their ability to pursue goals over time (see further, Bennett and Hacker, 2008, chapter 5). The role of emotions in caring about moral issues also explains the results of Greene's experiments (Greene *et al.*, 2001; Greene, 2002): he found in fMRI investigations that people, when they are engaged in moral (and, what Greene calls, personal) dilemmas, 'use' more emotion-related brain areas (such as the ventromedial prefrontal cortex) compared to situations in which they are engaged in nonmoral (and impersonal) dilemmas. But these experimental results do not show that their moral judgments are 'driven' by emotion-related brain areas, as Greene believes.

Prinz (2007) has advanced the idea that we acquire moral norms by emotional conditioning. This explains, according to him, why moral rules (in contrast to conventions) have a special normative force. Although emotional conditioning may be involved during the early stages of child development, Prinz' model does not explain the transition from instinctive behaviour to intentional and moral behaviour, both during ontogenesis and evolution (see further below). The reason why Prinz' ideas are problematic is that he thinks that identifying the object of an emotion plays a role in the perception (and cause) of a moral emotion (see Prinz, 2007, chapter 2). This is problematic, for there is no such thing as perception of one's own emotion. Furthermore, being able to say what emotion one is feeling, need not depend on identifying its cause but does depend on identifying its object (and the object of an emotion is surely not the cause of the emotion, as Prinz believes).

8.6　The nativist model

The emotionist model assumes that the affect system in our brain causes moral judgments. The nativist model starts at the other end of the spectrum and is developed to explain the universality of moral judgments. This model assumes that our knowledge of right and wrong is to a certain extent innate. Proponents of this model use Chomsky's ideas (e.g. the brief discussion of moral cognition in Chomsky, 1988) and argue that humans are equipped with an innate moral faculty which enables them to acquire knowledge of right and wrong. This moral faculty is, according to them, designed by natural selection and is comparable to the language faculty described by Chomsky (see, among others, Hauser, 2006; Mikhail, 2007). The moral faculty that humans possess is seen as a system that is present in, what Chomsky calls, the mind/brain. In terms of Chomsky's well-known principle and parameter model (Chomsky, 1980; 1988), this system is thought to consist of an abstract Universal Moral Grammar (UMG) incorporating a set of principles (also called rules, concepts). Some of these principles are fixed and rooted in our genetic endowment; others are characterized as parameters which obtain a value as the result of learning. This learning process is, however, according to proponents of the nativist model not comparable to how psychologists normally conceive of learning processes. They argue that an innate Acquisition Device somehow selects relevant data appropriate for the development of moral judgments. This selection process is seen as the core of the 'learning' process. The result is that the parameters appropriate for the environment in which the child is living are, as it were, switched on, resulting in the ability to develop moral judgments. Since all humans possess an innate moral faculty, acquiring the native moral system (appropriate for the environment in which a child is living) requires little to no instruction: experience with the native morality only sets the parameters left unspecified by Universal Moral Grammar (resulting in a specific moral system). Because UMG is unique for the human species, other animals do not acquire a moral system. And just as brain injury may lead to selective deficits in the language faculty (e.g. aphasia), it is argued that damage to mind/brain systems involved in moral judgments results in selective deficits in moral judgments (e.g. psychopathy).

Empirical evidence for the presence of UMG in the mind/brain is, however, not available. Yet proponents have argued that there are data that point in the direction of UMG. Do these data support the idea of UMG? First, it is argued that 'the intuitive jurisprudence of young children is complex and exhibits many characteristics of a well-developed legal code' (Mikhail, 2007, p. 143). This claim is based on the observation that 4- to 5-year-old children use words such as 'intent', 'purpose' and 'ought', and are able to distinguish moral violations (e.g. theft) from violations of social conventions (e.g. wearing pyjamas to school). It is, however, not an argument in favour of UMG, for it only shows that children (have learnt to) use these words for distinguishing intentional behaviour from other types of behaviour. They understand that, in the case of the (purposeful) violation of a norm (hence they understand the difference between good and bad) children could have refrained from action. Second, every natural language seems to have words or phrases to express deontic concepts (e.g. obligatory, permissible and forbidden). This is, however, not an argument is favour of UMG, for it only shows that these concepts are used in different languages. Third, prohibitions of murder, rape and other forms of aggression seem to be universal. Again, this observation is not an argument in favour of UMG for it only shows that standards of moral conduct ('one should not harm another person') are recognized in most cultures as objective or correct standards for behaviour. It tells us nothing about the innateness of UMG.

Proponents of the nativist model have also discussed two general arguments, derived from Chomsky's ideas about the Language Faculty, as

evidence for the existence of UMG. It is said that the explanatory framework
relies on these arguments. First, just as in the case of acquiring the ability to
speak and understand language, it is believed that children learn moral
judgments fast. It is insisted that this is remarkable because of the poverty
of the moral stimuli. However, just as in the case of acquiring a language,
there is no empirical evidence in favour of the argument from the poverty of
the moral stimulus. The problem here is that proponents of the nativist
model confuse acquiring knowledge of the difference between right and
wrong with acquiring knowledge about models from ethics (utilitarianism,
etc.). For knowledge of the models of Rawls and others is not necessary for
understanding the difference between right and wrong. Second, the 'fact'
that children acquire a moral grammar of great complexity suggests,
according to some investigators, that UMG is designed as a hypothesis-
forming or computational system. Otherwise it is, according to them,
inexplicable how children are able to develop moral judgments for problems
that they have not encountered before. The problem here is that this alleged
'fact' is only puzzling or mysterious if it is assumed that UMG is a system in
the mind/brain. For this problem does not arise if we note that children have
an innate ability to acquire moral behaviour. Some abilities (for example,
instinctive responses) are rigid; other abilities are plastic (mastery of a
language or of the standards for moral conduct). If knowledge is involved,
as is the case in moral behaviour, then the ability to develop moral judgments is open-ended (see further Baker and Hacker, 1984; Hacker, 1990a).
Hence it is unsurprising that someone who understands the difference
between right and wrong is able to understand many moral problems,
including new problems that he or she has not experienced before.

What is the alternative framework for understanding the data discussed
by proponents of the nativist model? Just as it is better (since it is
conceptually coherent; see Chapter 6) to say that humans have an innate
ability to acquire a language, it is better to say that humans have an
innate ability to acquire (and understand, use) standards of moral conduct. This explains why the development of moral behaviour is preceded
by or goes hand in hand with the development of volitional and intentional behaviour out of instinctive behaviour. This also clarifies why only
humans and not other animals, as Darwin noted, can be held to be
morally responsible for their deeds, for they can (be asked to) justify what
they have done or intend to do because they are responsive to reason.
I discuss two clarifications.

First, according to the nativist model, animals lack an innate moral
faculty (consisting of UMG as a component in the mind/brain) and do

not, therefore, develop a sense of 'oughtness' or what is also called a sense of obligation. Yet according to the alternative view elaborated here, there is no such thing as Universal Moral Grammar present somewhere in the mind/brain. Children have an innate ability to acquire a moral language and, hence, acquire standards for moral behaviour. For their moral behaviour and reasoning is an extension of the development of intentional behaviour, i.e. the ability to give reasons for what they are doing. It starts therefore when children are able to give reasons for what they are doing. Some of these reasons invoke principles for moral conduct or values, i.e. 'One should not lie' or 'Stealing is wrong.' If they are recognized as universal moral rules or objectives for conduct, then children will use these standards for evaluating their own and others' behaviour. This does not imply, of course, that children always use these rules as standards for conduct (for children, displaying intentional behaviour, can be insincere, etc.). Second, the main difference between humans and other animals is not that animals lack Universal Moral Grammar but the innate ability to acquire a language. Animals are not able to give reasons for their goal-directed behaviour and therefore develop intentional behaviour only in an attenuated sense. Hence it is possible that, for example, dogs adhere to some of the rules that we set for them, but they do not develop then a moral sense of obligation (see Hanfling, 2003; Strawson, 2008 [1974]). It does not make sense in the case of misbehaviour of the dog to blame the dog for misbehaving, for a dog, lacking the ability to acquire a language, cannot ward off blame by means of excuses as children can. This explains why most of us do not experience the moral attitude of resentment or indignation when a dog misbehaves. Of course, we may be angry with the dog, but the anger is then not an example of a moral emotion. By contrast, we are resentful with someone who does not keep a promise and this is an example of a moral emotion, because the anger tells us that we are upset about the fact that someone is insincere and not trustworthy.

8.7 The evolution of fair sharing

The model I advance states that only humans act on norms because they are language-using creatures. Children acquire this ability when they pick up words and learn a moral language. But this does not mean that children learn to use words such as 'good' and 'bad' as parts of an abstract moral system. Children learn the use of these words when they observe or hear about acts that are said to be right or wrong, or when a parent evaluates their behaviour. Suppose that a child takes the toy of his brother or sister, then the parent may say 'Bad boy!' or 'Don't do that!' and, hence, expresses his or her disproval. The child then learns the difference between right and wrong and begins to understand that certain behaviours are (evaluated as) good or bad. This involves an emotive element, for the sentences are expressed together with smiles in the case of approval, or frowns and physical prevention in the case of disproval. And if a child does something that it ought not to do, the caregiver might add that the child should be ashamed of itself (and will perhaps punish or shame the child). Hence the child acquires also a sense of obligation and feelings of shame, remorse and guilt. Acquiring moral behaviour goes together with learning to care about moral issues.

Caring about others and consideration for others are moral virtues which children learn within the family and later the group in which they are living. Teasing or bullying your brother is bad and the caregiver will add a reason: 'It will upset your brother.' Helping and caring for a sick or distressed brother is good and contributes to the good of him. Caregivers stimulate affection and sympathy by praise. Hence acquiring the difference between right and wrong with regard to social behaviour is also the beginning of acquiring the altruistic virtues (kindness, benevolence), respect for property ('You should not take your sister's toy') and a sense of fairness (when children learn to share food in a family and are confronted with solving disputes among children). Parents often ask in such situations: 'How would you like it if another person harmed you or took your property?', and this is a way of directing the attention of children to the needs, wishes and plans of another. And by calling behaviours 'right' or 'wrong', the parent introduces also the possibility of explaining (giving reasons) why these behaviours are right or wrong. They also encourage children to look for reasons for their behaviour. 'Why is it reprehensible to pull the cat's tail?'; 'Why should you share toys with other children?'; 'If someone did this to you, how should you react?' Hence reasoning is part of moral education and internalizing norms of moral conduct is, therefore, not only a form of conditioning; it is also acquiring the ability to give reasons for acts through answering questions asked by caregivers. Through answering such questions children learn to account for what they are doing and become responsible for their deeds. It helps children to explain their intended acts and justify acts through (backward- and forward-looking) reasons. And if they have learnt to give reasons, they know why others may blame them when they do something that they were told not to do ('Why are you doing that when I told you not to?'). Older children learn to ward off blame by means of excuses ('I couldn't help it') or by saying that 'there was no opportunity for doing anything else'. Still older children learn to respond to the requests of parents by saying: 'I do not want to.' Hence acquiring knowledge results in cases where children decide to do something contrary to what a caregiver has asked or told the child to do. In such cases the caregiver will explain what the reasons are for asking the child to do something and why he or she expects an appropriate reason for doing something else in the given context. These reasons are, in the context of a family and the group, especially important with regard to possible conflicts between self-interest and other-regarding behaviour. These conflicts have also played a role during the evolution of moral behaviour. I shall only discuss the evolution of the ability to act on norms in the context of hunting large game.

The ability to act on norms is an interesting topic for inclusive fitness theory, for investigating its evolution will provide us with insight into the evolution of typical human forms of cooperation. One possibility is that acting on norms evolved when hominids started to hunt large game, for it resulted in the use of *public goods* in hunter-gatherer societies. The use of public goods is believed to have been constitutive for the formation of these societies (see Hawkes, 1993). Hominids started to hunt big game subsequent to the use of advanced tools in cooperative hunting. Other animals also hunt in groups, but a major innovation in the early hominids was the use of spears and others tools. They also started then to transport big game to a fixed location (what would become their home). Chimpanzees are also able to carry objects but are far less skilled and do not bring these objects to a certain location. They display communal feeding (but do not share food) only at the location where they are foraging.

Sharing a public good in the group enhanced inclusive fitness, but also created opportunities for cheating, for the public good can be consumed by those who do not pay the acquisition costs. Hence one can imagine that 'solving' the problem of free-riders required an extension of the communicative behaviours of the humans. I suggest that it resulted in new forms of complex linguistic behaviour, namely the ability *to act on* norms. This is an example of a complex skill, for it presupposes that humans understand the difference between right and wrong and, hence, comprehend that doing something can be desirable or obligatory given certain reasons or values. It requires that humans can reason and deliberate, and can weigh the pros and cons of facts that they know in the light of their desires, goals and values. It is interesting that sharing food (as the result of the ability to act on norms) is one of the major characteristics of hunter-gatherer societies (Hill *et al.*, 2011).

Children develop the ability to act on norms during the so-called 5–7 shift (Campbell, 2006), i.e. the beginning of the adrenarche (the production of androgens by the adrenal glands). It corresponds to the shift in their sustenance from the nuclear family to the group (see also Haig, 2010). Suppose that a child is the oldest one in the family. Adrenarche occurs at the time that the mother is pregnant with her third child while the second child is weaning. It is thought that the adrenarche in humans evolved because it enabled children to participate in the group. Úbeda's model (2008; Úbeda and Gardner, 2010; 2011) predicts here a reversal in imprinted gene expression (see also Chapter 3, section 3.6).

There are two possible reasons for the effects of imprinted genes during the 5–7 shift. First, anthropologists have shown that in hunter-gatherer societies, males contribute more to food provisioning in the group than females (in terms of calories; see Ember, 1978). This raises the possibility that alleles that come from the mother favour more selfish behaviour than genes from the father (just as alleles that come from the father promote more selfish behaviour when children are largely dependent on maternal investments). Úbeda (2008) has argued that this pattern explains why children with Prader–Willi syndrome (PWS), lacking the expression of patrigenes, hardly suck but become hyperphagic after about five years, when they utilize paternal resources. Second, because males in the group are more related than females, it is likely that paternally expressed genes contribute to cooperative behaviour, while maternally derived genes favour free-riding. This may explain another aspect of the behaviour of children with PWS, namely their 'food hoarding and stealing'. Interestingly, there is evidence that patrigenes are involved in adrenarche (see Siemensma et al., 2011). This raises the possibility that adrenal androgens, through affecting the development and activities of brain structures such as the amygdala, hippocampus and the insula (Campbell, 2006; 2011), are involved in moral development. Yet how do patrigenes affect moral development? Children develop moral behaviour when they acquire the capacity to care about right and wrong. And when they care about moral issues, then they experience emotions, for one feels no emotions about things concerning which one is indifferent, and one does not pursue goals efficiently unless one cares about achieving them. Applied to the example of sharing food, it means that children are resentful with someone who does not share food (and hence violates a rule). The anger they experience when someone violates a norm motivates them to punish cheaters. This can be seen as a part of an evolved adaptation, because experiencing moral emotions (as part of complex, linguistic moral behaviour) helped them to pursue shared

Table 8.1: Different interests and possible effects of patrigenes and matrigenes (described in catchwords) during the transition from family to group life (the 5–7 shift).

Patrigenes	Matrigenes
Sharing food	Hoarding and stealing food
Egalitarianism	Egocentrism
Moral sense, social norms	Defecting, lying

goals efficiently through collaborative action. I suggest therefore that patrigenes (involved in adrenarche), through affecting the development of brain structures such as the amygdala and hippocampus, contribute in this way to the development of the moral emotion required for a fair sharing of food, whereas matrigenes benefit during the 5–7 shift by stealing from the communal pot (see Table 8.1). Interestingly, the behavioural profile of children with PWS is described as egocentric: they 'argue, lie, manipulate and confabulate to change rules' and their social judgment is poor, reminding clinicians of children with pervasive developmental disorder (Cataletto *et al.*, 2011).

8.8 Conclusion

Darwin argued that only humans evolved into moral beings. I have extended his arguments through discussing the role of language in moral development. The main difference between humans and other animals is that only humans use a language. The use of language enables us to follow rules: they constitute reasons for doing things and inform actions within rule-governed practices. I have argued that the ontogenesis and evolution of rule-governed behaviour is an example of the development of a complex, linguistic skill. Moral behaviour evolved as an extension of this skill.

I have contrasted the elaborations of Darwin's ideas with the emotionist and nativist models in moral psychology. The emotionist model tries to explain the problematic relation between moral cognition and emotion through postulating different systems in the brain. It assumes that emotions precede or cause moral judgments. An important reason why the emotionist model is incoherent is that the model does not differentiate the cause and object of an emotion and misinterprets therefore the relation between emotion and cognition. The nativist model assumes that humans suddenly (i.e. as the result of a mutation) acquired innate knowledge of right and wrong. I have argued that humans are not only unique because they possess an innate moral grammar, but because they have the second-order ability to acquire a language. This ability enables humans to acquire moral behaviour as an extension of intentional behaviour. I have used the evolution of the incest taboo to demonstrate the falsity of the emotionist and nativist models and have discussed the evolution of fair sharing as an illustration of the alternative model.

Language differentiates humans from the other animals. Consequent on becoming language-using creatures, humans became culture-creating and self-conscious creatures who evolved into moral beings when they understood what actions are, what right and wrong is and when they understood that it is possible to make amends. They became rational animals that are responsive to reasons and can act for reasons, including norms and values. This explains why Darwin correctly argued that only humans can be ranked as moral beings.

CHAPTER 9

Epilogue

I have discussed in this book a general framework integrating Hamilton's inclusive fitness theory and the neo-Aristotelian conception of animate nature. I have also discussed the advantages of this framework through comparing it to the alternative Cartesian conception extended with modern evolutionary theory. The example of genomic imprinting was discussed in order to show how the framework can be used for developing hypotheses. Since neo-Aristotelians emphasize that mastery of a language is the mark of the human mind, we can expect that the Aristotelian conception leads to new predictions when imprinted genes affect typically human communicative behaviours (see also Smit, 2010a; 2010d). I have discussed how the Aristotelian conception can be used here for generating hypotheses and why it sheds new light on the evolution and ontogenesis of the animal and human mind. I repeat here the reasons why I believe that crypto-Cartesians have it wrong, which are primarily conceptual ones. The Cartesian conception leads to misconceptions about the relation between the brain and the mind and, hence, to misunderstandings about possible effects of genes on the development of the brain and mind.

Extending inclusive fitness theory with the neo-Aristotelian conception has another advantage: it enables us to understand the evolutionary transition from animal to human societies. Hamilton's inclusive fitness theory provides us with an overarching conceptual framework for explaining the major evolutionary transitions in animate nature. The framework elucidates how cooperation between lower-level units has resulted in higher-level creatures as the result of either direct or indirect fitness benefits. I have argued that by invoking the neo-Aristotelian conception, we can understand how the transition from animal to human societies was accomplished. Linguistic behaviour evolved as a new form of communicative behaviour (and replaced, in part, older, animal forms of communication), enabled humans to develop new skills and extended therefore behavioural flexibility. It resulted in the extreme forms of division of labour characterizing human societies. Direct fitness benefits (combining different functions) explain the advantages of these divisions of labour, but it is likely that indirect fitness benefits have been important too during the early stages of language evolution when humans lived in small groups. Thus inclusive fitness theory explains at least the early stages of the sixth transition that later culminated in the typical human, rational powers of the intellect and will. It helps us to understand why only humans became self-conscious, culture-creating beings, and why natural selection was later extended with cultural selection.

Integrating inclusive fitness theory and the neo-Aristotelian conception enables us to explain social behaviour in both the subcellular and supracellular world *and* the emergence of the unique features of human nature. It also helps us to understand why previous attempts were less successful, because they used (variants of) the incoherent, Cartesian conception. I have discussed in this book the coherent alternative, but realize that it may still create confusion because I have not criticized the practice among many evolutionary theorists of anthropomorphizing animate nature. For example, when evolutionary theorists explain the fitness effects of the four possible interactions between an actor and a recipient, they use the 'technical terms' selfish and altruistic behaviour, mutual benefit and spite. Moreover, when they apply these terms to the 'behaviour' of genes, cell organelles, cells or organisms, they are inclined to say that, for example, cell organelles behave 'selfishly' or that a cell acts 'altruistically'. Applying the same technical terms to cell organelles, cells, organisms, etc. is done in order to highlight that the same principles of social evolution are applicable at several levels of organization. And because ultimate causal explanations are complementary to the proximate causal explanations developed by chemists, cell biologists, social scientists, etc., the use of these 'technical terms' does not mean that there are no differences between – say – chemical reactions, activities of cells, behaviours of animals and motivated actions of humans. Evolutionary theorists emphasize that these differences are explicable in terms of proximate mechanisms (or motivations, intentions, etc.). Moreover, by calling these terms 'technical' evolutionary theorists emphasize that it is not their intention to anthropomorphize animate nature. I have assumed that readers understand the distinction between technical and ordinary terms. I have also assumed that it is clear that the 'behaviour' displayed by cell organelles and cells does not license us to apply predicates such as 'egoism' or 'coercion' to these creatures. I have tried to avoid misunderstandings by emphasizing the distinction between empirical and conceptual truths, and by discussing the principles of social evolution and the Aristotelian conception as separate topics. Nevertheless I realize there are still many opportunities for misunderstandings.

The framework I have discussed clarifies what the unique features of human nature are and how these can be explained. It deviates from ethology and sociobiology (and disciplines that flourished later like evolutionary medicine and psychology). Ethologists argued (correctly) that we can understand the animal mind through observing its behavioural manifestations (see Smit, 1989). They thought that we could investigate the human mind by using the very same methods. Yet their approach left unexplained how the unique human mind evolved as the result of language evolution and, hence, how we can explain essential (behavioural) differences between humans and other animals (and what the methodological consequences are for studying these differences). Sociobiologists argued (correctly) that the principles of social evolution are applicable to all levels of organization, including human societies. Inclusive fitness theory was therefore believed to be capable of solving the major problems in the social sciences and humanities. However, just as with ethology, sociobiology did not provide us with a satisfactory answer to the question of how evolutionary principles can be used to explain unique features of the human species, like behavioural flexibility, self-consciousness or acting for reasons. Even worse: by calling the evolution of self-consciousness a mystery or by saying that knowledge of Universal (Moral) Grammar is inaccessible to scientific investigation, it was unclear why social scientists should pay attention to explanations put forward by evolutionary theorists. With the benefit of hindsight, we can easily explain why these attempts were doomed to fail: these theorists used variants of the incoherent Cartesian conception of human nature. Mutatis mutandis, we can expect that the Aristotelian framework extended with Hamilton's theory is better suited as a promising starting point for studying the unique features of human nature.

Because I have advanced the Aristotelian conception instead of the Cartesian one, there is an important difference between my proposal and older proposals. Neo-Aristotelians argue that mastery of a language is the mark of the mind. And because they argue that language evolved from simple to more complex linguistic behaviour, they argue that the horizon of human thought is not fixed (as Kant and others believed), but determined by what humans can express in non-verbal and linguistic behaviour. What humans could do with words expanded during the course of evolution and gradually resulted in a gap between humans and other animals.

The most obvious example is that only humans are able to form intentions, and reasoned behaviour requires the use of a language. The origin of
the ability to act for reasons is, in part, explicable in terms of inclusive
fitness theory, but this does not mean that there are no essential differences
between animal and human behaviour, because linguistic behaviour
evolved as a new form of behaviour resulting in unique rational powers.

The Aristotelian framework extended with Hamilton's theory combines
theoretical and conceptual truths. It is therefore capable of generating new
and testable hypotheses about the social evolution of human nature. I have
discussed some examples in this book and note here that there is ample
room for improving these hypotheses and for coming up with new and
better ones. Language evolution and its consequences for the social evolution of human nature require more empirical and conceptual attention
than I have given them in this book. I expect that we will learn more about
the evolution of human nature as soon as we learn more about the genes
that differentiate us from the other apes. Yet to explain the effects of these
genes, we have to elaborate the neo-Aristotelian conception in order to get
a better understanding of the evolution of linguistic behaviour. I have
distinguished simple (asking, demanding, etc.) and complex (thinking,
imagining, etc.) forms of linguistic behaviour, but it is obvious that it is
more interesting to subdivide the stages of the transition from animal to
linguistic behaviour by using other terms than 'simple' and 'complex'. This
is an interesting topic for future studies, the more so because such studies
will shed new light on old philosophical problems.

The example of genomic imprinting and the possible effects of matrigenes and patrigenes on brain and behavioural development were discussed
as an illustration. The reason why I have chosen this phenomenon was that
I started to think about integrating the Aristotelian conception and inclusive fitness theory after I discovered intragenomic conflicts (in the context
of early child development) and noticed that imprinted genes affect brain
development. The phenomenon of genomic imprinting is an interesting
case study for two reasons. First, it is an example demonstrating how a
chemical phenomenon can be explained by an appeal to arguments based
on inclusive fitness theory. Molecular biology clarifies how attaching or
removing methyl groups to the bases in the DNA results in silencing or
activating genes; evolutionary biology explains why methylation of the
DNA depends on the sex of the parent (and why there is probably a
reversal in the expression of imprinted genes in humans). Hence proximate
and ultimate causal explanations are clearly integrated here into one
explanatory framework. Second, the effects of imprinted genes on brain

development raise the question if and how imprinted genes affect the development of human nature. I have discussed some hypotheses, but emphasize that they are speculative hypotheses, since it is still unclear how imprinted genes are involved in the evolution of the human mind (e.g. moral behaviour and the early forms of typical human communicative behaviour). The reason why there is still uncertainty is that it is unclear what the precise effects are of human variations of imprinted genes. Many imprinted genes were selected before humans evolved as a separate species. For instance, studies of the evolution of the 15q11-q13 imprinted cluster involved in PWS and AS have shown that this cluster originated about 105 million to 180 million years ago (Rapkins *et al.*, 2006). Since then the region expanded through insertions (three genes have been added 90 million to 105 million years ago) and it has undergone some rearrangements afterwards. There is also evidence that there are differences between the region in humans and the homologous region in others animals. For example, the region harbours only in humans the patrigene NPAP1, which is expressed in neurons. Hence the 15q11-q13 cluster appears to be an unstable region still subject to selection. It is therefore possible, but not proven, that imprinted genes in the 15q11-q13 cluster contribute to the development of uniquely human forms of communicative behaviour. We have to await the results of future studies to see whether the hypotheses that I have advanced hold.

When empirical evidence showed that imprinted genes affect brain development, it was interesting for someone with a background in analytic philosophy to think about integrating inclusive fitness theory and the neo-Aristotelian conception of human nature. In a previous book (Smit, 2010c) I have discussed some implications of the framework integrating inclusive fitness theory and the Aristotelian conception for studying diseases and psychopathology (such as autism and anorexia nervosa). It is interesting to reconsider other topics in terms of the neo-Aristotelian framework extended with evolutionary theories.

I have argued that philosophical problems (like the mind/body problem) are conceptual problems generated by misconceptions. They can (and must) be resolved through conceptual investigations. The insight that philosophical problems are conceptual ones is the result of the so-called linguistic turn in philosophy. It has helped us to get a clearer picture of what the interesting empirical and conceptual issues are in biology, medicine and psychology. The linguistic turn also helped us to answer the question of how we can integrate evolutionary theory and the Aristotelian conception of animate nature. I expect that elaborations of the neo-Aristotelian conception extended with Hamilton's theory are going to transform the ideas of scientists and philosophers about human nature in the near future. Yet I have also noticed that many scientists and philosophers prefer another framework than the one I have advanced in *The social evolution of human nature*. Time will tell whether this book contributes to the spread of theoretical and conceptual insights and whether it leads to a change in the way scientists and philosophers think about human nature.

References

Allen, N. D., Logan, K., Lally, G., Drage, D. J., Norris, M. L. and Keverne, E. B. (1995) Distribution of parthenogenetic cells in the mouse brain and their influence on brain development and behaviour. *Proceedings of the National Academy of Sciences USA* 92: 10782–10786.

Aristotle (2002 [1968]) *De Anima*, Clarendon Aristotle Series, translated by W. D. Hamlyn. Oxford University Press.

Arnhart, L. (2005) The incest taboo as Darwinian natural right, in A. P. Wolf and W. D. Durham (eds.), *Inbreeding, incest and the incest taboo*. Stanford University Press, 190–217.

Axelrod, R. and Hamilton, W. D. (1981) The evolution of cooperation. *Science* 211: 1390–1396.

Ayala, F. J. (1970) Teleological explanations in evolutionary biology. *Philosophy of Science* 37: 1–15.

(2007) Darwin's greatest discovery: design without designer. *Proceedings of the National Academy of Sciences USA* 104: 8567–8573.

Badcock, C. and Crespi, B. (2006) Imbalanced genomic imprinting in brain development: an evolutionary basis for the aetiology of autism. *Journal of Evolutionary Biology* 19: 1007–1032.

(2008) Psychosis and autism as diametrical disorders of the social brain. *Behavioral and Brain Sciences* 31: 241–261.

Baker, G. P. and Hacker, P. M. S. (1984) *Language, sense and nonsense: a critical investigation into modern theories of language*. Oxford: Blackwell.

(2005 [1983]) *Wittgenstein: meaning and understanding*, second edition, revised by P. M. S. Hacker. Oxford: Blackwell.

(2009 [1985]) *Wittgenstein: rules, grammar and necessity*, second edition, revised by P. M. S. Hacker. Oxford: Basil Blackwell.

Balaban, E. (1997) Changes in multiple brain regions underlie species differences in a complex, congenital behaviour. *Proceedings of the National Academy of Sciences USA* 94: 2001–2006.

Baldwin, J. M. (1892) Origin of volition in childhood. *Science* 20: 286–287.

(1896) A new factor in evolution. *American Naturalist* 30: 441–451, 536–553.

Barnes, J. (1984) *The complete works of Aristotle*, vol. 1. Princeton University Press.

Barnes, K. C., Armelagos, G. J. and Morreale, S. C. (1999) Darwinian medicine and the emergence of allergy, in W. R. Trevathan, E. O. Smith and J. J. McKenna (eds.), *Evolutionary medicine*. Oxford University Press, 209–243.

Baron-Cohen, S. (1995) *Mindblindness: an essay on autism and theory of mind*. Cambridge, MA: MIT Press.

  (2003) *The essential difference: the truth about the male and female brain*. New York: Basic Books.

Bennett, M. R. and Hacker, P. M. S. (2003) *Philosophical foundations of neuroscience*. Oxford: Blackwell.

  (2008) *History of cognitive neuroscience*. Chichester: Wiley-Blackwell.

  (2011) Criminal law as it pertains to patients suffering from psychiatric disease. *Bioethical Inquiry* 8: 845–858.

Bittel, D. C. and Butler, M. G. (2005) Prader–Willi syndrome: clinical genetics, cytogenetics and molecular biology. *Expert Reviews in Molecular Medicine* 7: 1–20.

Blair, J., Mitchell, D. and Blair, K. (2005) *The psychopath: emotion and the brain*. Oxford: Blackwell.

Blass, E. M. and Teicher, M. H. (1980) Suckling. *Science* 210: 15–22.

Blurton Jones, N. G. and da Costa, E. (1987) A suggested adaptive value of toddler night walking: delaying the birth of the next sibling. *Ethology and Sociobiology* 8: 135–142.

Bogin, B. (1997) Evolutionary hypotheses for human childhood. *Yearbook of Physical Anthropology* 40: 63–89.

Bogin, B. and Smith, H. (1996) Evolution of the human life cycle. *American Journal of Human Biology* 8: 703–716.

Bonner, J. T. (1974) *On development: the biology of form*. Cambridge, MA: Harvard University Press.

  (1980) *The evolution of culture in animals*. Princeton University Press.

Boomsma, J. J. (2009) Lifetime monogamy and the evolution of eusociality. *Philosophical Transactions of the Royal Society B* 364: 3191–3207.

Bourke, A. F. G. (2011) *Principles of social evolution*. Oxford University Press.

Boyd, R., Richerson, P. J. and Henrich, J. (2011) The cultural niche: why social learning is essential for human adaptation. *Proceedings of the National Academy of Sciences USA* 108: 10918–10925.

Bradley, F. H. (1893) *Appearance and reality*. New York: Macmillan.

Breinl, F. and Haurowitz, F. (1930) Chemische Untersuchungen des Präzipitates aus Hämoglobin und anti-Hämoglobin Serum und Bemerkungen über die Natur der Antikörper. *Hoppe-Seyler's Zeitschrift für Physiologische Chemie* 192: 45–57.

Brodin, A. and Ekman, J. (1994) Benefits of food hoarding. *Nature* 372: 510.

Brown, W. M. and Consedine, N. S. (2004) Just how happy is the happy puppet? An emotion signalling and kinship theory perspective on the behavioral phenotype of children with Angelman syndrome. *Medical Hypotheses* 63: 377–385.

Buiting, K. (2010) Prader–Willi syndrome and Angelman syndrome. *American Journal of Medical Genetics Part C* 154C: 365–376.

Burkart, J. M., Hrdy, S. B. and van Schaik, C. P. (2009) Cooperative breeding and human cognitive evolution. *Evolutionary Anthropology* 18: 175–186.

Burnet, F. M. (1957) A modification of Jerne's theory of antibody production using the concept of clonal selection. *Australian Journal of Science* 20: 67–69.

Burt, A. and Trivers, R. (2006) *Genes in conflict: the biology of selfish genetic elements.* Cambridge, MA: Belknap Press of Harvard University Press.

Bush, J. R. and Wevrick, R. (2012) Loss of the Prader–Willi obesity syndrome protein necdin promotes adipogenesis. *Gene* 497: 45–51.

Buss, D. (1999) *Evolutionary psychology: the new science of the mind.* Boston, MA: Allyn & Bacon.

Buss, L. W. (1987) *The evolution of individuality.* Princeton University Press.

Butler, M. G. (1990) Prader–Willi syndrome: current understanding of cause and diagnosis. *American Journal of Medical Genetics* 35: 319–332.

Campbell, B. (2006) Adrenarche and the evolution of human life history. *American Journal of Human Biology* 18: 569–589.

(2011) Adrenarche and middle childhood. *Human Nature* 22: 327–349.

Candlish, S. (1995) Kinästhetische Empfindungen und epistemische Phantasie, in E. von Savigny and O. R. Scholz (eds.), *Wittgenstein über die Seele.* Frankfurt am Main: Suhrkamp, 159–193.

Caspi, A., McClay, J., Moffit, T. E., Mill, J., Martin, J., Craig, I. W., Taylor, A. and Poulton, R. (2002) Role of genotype in the cycle of violence in maltreated children. *Science* 297: 851–853.

Cataletto, M., Angulo, M., Hertz, G. and Whitman, B. (2011) Prader–Willi syndrome: a primer for clinicians. *International Journal of Pediatric Endocrinology* 18 Oct.: 12.

Chapais, B. (2008) *Primeval kinship.* Cambridge, MA: Harvard University Press.

Charrier, C., Joshi, K., Coutinho-Budd, J., Kim, J-I., Lambert, N., de Marchena, J., Jin, W-L., Vanderhaeghen, P., Gosh, A., Sassa, T. and Polleux, F. (2012) Inhibition of SRGAP2 function by its human-specific paralogs induces neoteny during spine maturation. *Cell* 149: 923–935.

Chomsky, N. (1959) A review of B. F. Skinner's *Verbal Behaviour. Language* 35: 26–58.

(1980) *Rules and representations.* Oxford: Blackwell.

(1988) *Language and problems of knowledge: the Managua lectures.* Cambridge, MA: MIT Press.

Clayton, J. and Laan, L. (2003) Angelman syndrome: a review of the clinical and genetic aspects. *Journal of Medical Genetics* 40: 87–95.

Clutton-Brock, T. (2009) Cooperation between non-kin in animal societies. *Nature* 462: 51–57.

Cook, J. W. (1969) Human beings, in P. Winch (ed.), *Studies in the philosophy of Wittgenstein.* London: Routledge & Kegan Paul, 117–151.

Crick, F. (1995) *The astonishing hypothesis: the scientific search for the soul*. New York: Touchstone.

Crick, F. and Koch, C. (2002) The problem of consciousness. *Scientific American* 12: 10–17, updated from the September 1992 issue.

Csibra, G. and Gergely, G. (2006) Social learning and social cognition: the case for pedagogy, in Y. Munakata and M. H. Johnson (eds.), *Processes of change in brain and cognitive development*. Oxford University Press, 249–274.

Csibra, G. and Gergely, G. (2009) Natural pedagogy. *Trends in Cognitive Science* 13: 148–153.

Curio, E. (1973) Towards a methodology of teleonomy. *Experientia* 29: 1045–1058.

Damasio, A. R. (1996) The somatic marker hypothesis and the possible functions of the prefrontal cortex. *Philosophical Transactions of the Royal Society B* 354: 1413–1420.

Darwin, C. (2003 [1871]) *The descent of man, and selection in relation to sex*, with an introduction by Richard Dawkins. London: Gibson Square Books Ltd.; first published in 1871 by John Murray.

(1877) A biographical sketch of an infant. *Mind: A Quarterly Review of Psychology and Philosophy* 2: 285–294.

(1993 [1958]) *The autobiography*, edited by N. Barlow. New York: W. H. Norton & Company.

Darwin, F. (ed.) (1887) *The life and letters of Charles Darwin, including an autobiographical chapter*, 3 vols. London: John Murray.

Dawkins, R. (1976) *The selfish gene*. Oxford University Press.

(1982) *The extended phenotype: the gene as the unit of selection*. San Francisco: W. H. Freeman.

Dencker, S. J., Johansson, G. and Milson, I. (1992) Quantification of naturally occurring benzodiazepine-like substances in human breast milk. *Psychopharmacology* 107: 69–72.

Dennett, D. (1991) *Darwin's dangerous idea*. New York: Simon & Schuster.

Dennis, M. Y., Nuttle, X., Sudmant, P. H., Antonacci, F., Graves, T. A., Nefedov, M., Rosenfeld, J. A., Sajjadian, S., Malig, M., Kotkiewicz, H., Curry, C. J., Shafer, S., Shaffer, L. G., de Jong, P. J., Wilson, R. K. and Eichler, E. E. (2012) Evolution of human-specific neural SRGAP2 genes by incomplete segmental duplication. *Cell* 149: 912–922.

Descartes, R. (1985a) Discourse on the method, in *The philosophical writings of Descartes*, vol. 1, translated by J. Cottingham, R. Stroothoff and D. Murdoch. Cambridge University Press, 111–151.

(1985b) Principles of philosophy, in *The philosophical writings of Descartes*, vol. 1, translated by J. Cottingham, R. Stroothoff and D. Murdoch. Cambridge University Press, 193–291.

DeVeale, B., van der Kooy, D. and Babak, T. (2012) Critical evaluation of imprinted gene expression by RNA–Seq: a new perspective. *PLoS Genetics* 8(3): e1002600.

Didden, R., Korzilius, H., Duker, P. and Curfs, L. (2004) Communicative functioning in individuals with Angelman syndrome: a comparative study. *Disability and Rehabilitation* 26: 1263–1267.

Dilman, I. (1983) *Freud and human nature*. Oxford: Blackwell.

(1984) *Freud and the mind*. Oxford: Blackwell.

Ding, F., Li, H. H., Zhang, S., Solomon, N. M., Camper, S. A., Cohen, P. and Francke, U. (2008) SnoRNA Snord116 (Pwcr1/MBII-85) deletion causes growth deficiency and hyperphagia in mice. *PLoS Biology* 3(3): e1709.

Dondi, M., Simion, F. and Caltran, G. (1999) Can newborns discriminate between their own cry and the cry of another newborn infant? *Developmental Psychology* 35: 418–426.

Durham, W. H. (2005) Assessing gaps in Westermarck's theory, in A. P. Wolf and W. D. Durham (eds.), *Inbreeding, incest and the incest taboo*. Stanford University Press, 121–138.

Eggermann, T., Eggermann, K. and Schönherr, N. (2008) Growth retardation versus overgrowth: Silver–Russell syndrome is genetically opposite to Beckwith–Wiedemann syndrome. *Trends in Genetics* 24: 195–204.

Ember, C. (1978) Myths about hunter-gatherers. *Ethnology* 17: 439–448.

Enard, W. (2011) FOXP2 and the role of cortico-basal ganglia circuits in speech and language evolution. *Current Opinion in Neurobiology* 21: 415–424.

Erickson, M. T. (2005) Evolutionary thought and the current clinical understanding of incest, in A. P. Wolf and W. D. Durham (eds.), *Inbreeding, incest and the incest taboo*. Stanford University Press, 161–189.

Fischer, E. (1997) 'Dissolving' the 'problem of linguistic creativity'. *Philosophical Investigations* 20: 290–314.

(2003) Bogus mystery about linguistic competence. *Synthese* 135: 49–75.

Fisher, R. (1930) *The genetical theory of natural selection*. Oxford University Press.

Fisher, R. M., Cornwallis, C. K. and West, S. A (2013). Group formation, relatedness and the evolution of multicellularity. *Current Biology* 23: 1120–1125.

Fitch, W. T. (2010) *The evolution of language*. Cambridge University Press.

Flaxman, S. M. and Sherman, P. W. (2000) Morning sickness: a mechanism for protecting mother and embryo. *Quarterly Review of Biology* 75: 113–148.

Forbes, S. (2002) Pregnancy sickness and embryo quality. *Trends in Ecology and Evolution* 17: 115–120.

Foster, K. R. and Ratnieks, F. L. W. (2005) A new eusocial vertebrate? *Trends in Ecology and Evolution* 20: 363–364.

Fox, R. (1975) Primate kin and human kinship, in R. Fox (ed.), *Biosocial anthropology*. London: Malaby Press, 9–35.

(1980) *The red lamp of incest*. New York: E. P. Dutton.

(1989) *The search for society: quest for a biosocial science and morality*. New Brunswick: Rutgers University Press.

Frank, S. A. (1998) *Foundations of social evolution*. Princeton University Press.

(2003) Repression of competition and the evolution of cooperation. *Evolution* 57: 693–705.

Freud, S. (1950) *Totem and taboo*. New York: W. H. Norton. Published in German in 1913; translation by J. Strachey.

Gardner, A. (2009) Adaptation as organism design. *Biological Letters* 5: 861–864.

Gardner, A. and Grafen, A. (2009) Capturing the superorganism: a formal theory of group selection. *Journal of Evolutionary Biology* 22: 659–671.

Gardner, A. and West, S. A. (2010) Greenbeards. *Evolution* 64: 25–38.

Gardner, A., West, S. A. and Wild, G. (2011) The genetical theory of kin selection. *Journal of Evolutionary Biology* 24: 1020–1043.

Genevieve, D., Sanlaville, D., Faivre, L., Kottler, M. L., Jambou, M., Gosset, P., Boustani-Samara, D., Ointo, G., Ozilou, C., Abequillé, G., Munnich, A., Romana, S., Raoul, O., Cornier-Daire, V. and Vekemans, M. (2005) Paternal deletion of the GNAS imprinted locus (including GNASxl) in two girls presenting with severe pre- and post-natal growth retardation and intractable feeding difficulties. *European Journal of Human Genetics* 13: 1033–1039.

Gergely, G., Bekkering, H. and Király, I. (2002) Rational imitation in preverbal infants. *Nature* 415: 755.

Glackin, S. N. (2000) Animals, thoughts and concepts. *Synthese* 123: 35–64.

(2010) Universal grammar and the Baldwin effect: a hypothesis and some philosophical consequences. *Biology and Philosophy* 26: 201–222.

Glock, H-J. (2003) *Quine and Davidson on language, thought, and reality*. Cambridge University Press.

Gluckman, P. D. and Hanson, M. A. (2004) Living with the past: evolution, development, and patterns of disease. *Science* 305: 1733–1736.

Gluckman, P. D., Hanson, M. A., Bateson, P., Beedle, A. S., Law, C. M., Bhutta, Z. A., Anokhin, K. V., Bougnères, P., Chandak, G. R., Dasqupta, G., Smith, G. D., Ellison, P. T., Forrester, T. E., Gilbert, S. F., Jablonka, E., Kaplan, H., Prentice, A. M., Simpson, S. J., Uauy, R. and West-Eberhard, M. J. (2009) Towards a new developmental synthesis: adaptive developmental plasticity and human disease. *The Lancet* 373: 1654–1657.

Goldstone, A. P. (2004) Prader–Willi syndrome: advances in genetics, pathophysiology and treatment. *Trends in Endocrinology and Metabolism* 15: 12–20.

Gottlieb, G. (1972) *Development of species identification*. University of Chicago Press.

(1992) *Individual development and evolution*. Oxford University Press.

Grafen, A. (2003) Fisher the evolutionary biologist. *Journal of the Royal Statistical Society: Series D (The Statistician)* 52: 319–329.

(2006) Optimization of inclusive fitness. *Journal of Theoretical Biology* 238: 541–563.

(2008) The simplest formal argument for fitness optimisation. *Journal of Genetics* 87: 421–433.

(2009) Formalizing Darwinism and inclusive fitness theory. *Philosophical Transactions of the Royal Society B* 364: 3135–3141.

Gray, A. (1874) Scientific worthies: Charles Darwin. *Nature* 10 (4 June): 79–81.

Greene, J. D. (2002) Why are VMPFC patients more utilitarian? A dual-process theory of moral judgment explains. *Trends in Cognitive Sciences* 11, 322–323.

Greene, J. D., Sommerville, R. B., Nystrom, L. E., Darley, J. M. and Cohen, J. D. (2001) An fMRI investigation of emotional engagement in moral judgments. *Science* 293: 2105–2108.

Greer, P. L., Hanayama, R., Bloodgood, B. L., Mardinly, A. R., Lipton, D. M., Flavell, S. W., Kim, T. K., Griffith, E. C., Waldon, Z., Maehr, R., Ploegh, H. L., Chowdhury, S., Worley, P. F., Steen, J. and Greenberg, M. E. (2010) The Angelman syndrome protein Ube3A regulates synapse development by ubiquitinating Arc. *Cell* 140: 704–716.

Gregg, C., Zhang, J., Butler, J. E., Haig, D. and Dulac, C. (2010a) Sex-specific parent-of-origin allelic expression in the mouse brain. *Science* 329: 682–685.

Gregg, C., Zhang, J., Weissbourd, B., Luo, S., Schroth, G. P., Haig, D. and Dulac, C. (2010b) High-resolution analysis of parent-of-origin allelic expression in the mouse brain. *Science* 329: 643–648.

Hacker, P. M. S. (1986) *Insight and illusion: themes in the philosophy of Wittgenstein*, second, revised edition. Oxford: Clarendon Press.

(1987) *Appearance and reality: a philosophical investigation into perception and perceptual qualities*. Oxford: Blackwell.

(1990a) Chomsky's problems. *Language and Communication* 10: 127–148.

(1990b) *Wittgenstein: meaning and mind*. Oxford: Blackwell.

(1991) Seeing, representing and describing: an examination of David Marr's computational theory of vision, in J. Hyman (ed.), *Investigating psychology: sciences of the mind after Wittgenstein*. London: Routledge & Kegan Paul, 119–154.

(1995) Helmholtz' theory of perception: an investigation into its conceptual framework. *International Studies in the Philosophy of Science* 9: 199–214.

(1996a) *Wittgenstein: mind and will*. Oxford: Blackwell.

(1996b) *Wittgenstein's place in twentieth-century analytic philosophy*. Oxford: Blackwell.

(1999) *Wittgenstein on human nature*. New York: Routledge & Kegan Paul.

(2001) *Wittgenstein: connections and controversies*. Oxford: Clarendon Press.

(2002) Is there anything it is like to be a bat? *Philosophy* 77: 157–174.

(2004) The conceptual framework for the investigations of emotions. *International Review of Psychiatry* 16: 199–208.

(2005a) Of knowledge and of knowing that someone is in pain, in A. Pichler and S. Säätelä (eds.), *Wittgenstein: the philosopher and his works*. The Wittgenstein Archives at the University of Bergen, 123–156.

(2005b) Thought and action: a tribute to Stuart Hampshire. *Philosophy* 80: 175–197.

(2006) Passing by the naturalistic turn: on Quine's cul-de-sac. *Philosophy* 81: 231–253.

(2007) *Human nature: the categorial framework*. Oxford: Blackwell.

Hacker, P. M. S. (2012) Kant's transcendental deduction: a Wittgensteinian critique, in A. Marques and N. Venturinha (eds.), *Knowledge, language and mind: Wittgenstein's thought in progress*. Berlin: Walter de Gruyter, 11–35.

(2013) *The intellectual powers: a study of human nature*. Chichester: Wiley-Blackwell.

Haidt, J. (2001) The emotional dog and its rational tail: a social intuitionist approach to moral judgments. *Psychological Review* 108: 814–834.

(2007) The new synthesis in moral psychology. *Science* 316: 998–1002.

Haig, D. (1993) Genetic conflicts in human pregnancy. *Quarterly Review of Biology* 68: 495–532.

(1996) Placental hormones, genomic imprinting, and maternal-fetal communication. *Journal of Evolutionary Biology* 9: 357–380.

(1997) The social gene, in J. R. Krebs and N. B. Davies (eds.), *Behavioural ecology*, fourth edition. Oxford: Blackwell Scientific, 284–304.

(1999) Asymmetric relations: internal conflicts and the horror of incest. *Evolution and Human Behavior* 20: 83–98.

(2000) The kinship theory of genomic imprinting. *Annual Review of Ecology and Systematics* 31, 9–32.

(2002) *Genomic imprinting and kinship*. New Brunswick, NJ: Rutgers University Press.

(2004) Genomic imprinting and kinship: how good is the evidence? *Annual Review of Genetics* 38, 553–585.

(2006) Intrapersonal conflict, in M. Jones and A. C. Fabian (eds.), *Conflict*. Cambridge University Press, 8–22.

(2007). Weismann rules! OK? Epigenetics and the Lamarckian temptation. *Biology and Philosophy* 22: 415–428.

(2008) Huddling: brown fat, genomic imprinting and the warm inner glow. *Current Biology* 18: R172–R174.

(2010) Transfers and transitions: parent–offspring conflict, genomic imprinting, and the evolution of human life history. *Proceedings of the National Academy of Sciences USA* 107 (supplement 1): 1731–1735.

(2012) The strategic gene. *Biology and Philosophy* 27: 461–479.

Haig, D. and Grafen, A. (1991) Genetic scrambling as a defence against meiotic drive. *Journal of Theoretical Biology* 153: 531–558.

Haig, D. and Graham, C. (1991) Genomic imprinting and the strange case of the insulin-like growth factor-II receptor. *Cell* 64: 1045–1046.

Haig, D. and Úbeda, F. (2011) Genomic imprinting: an obsession with depilatory mice. *Current Biology* 21: R257–259.

Haig, D. and Wharton, R. (2003) Prader–Willi syndrome and the evolution of human childhood. *American Journal of Human Biology* 15: 320–329.

Haig, D. and Wilkins, J. F. (2000) Genomic imprinting, sibling solidarity and the logic of collective action. *Philosophical Transactions of the Royal Society B* 355: 1593–1597.

Hall, J. G. (1997) Genomic imprinting: nature and clinical relevance. *Annual Review of Medicine* 48: 35–44.

Hamburger, V., Wenger, E. and Oppenheim, R. (1966) Motility in the chick embryo in the absence of sensory input. *Journal of Experimental Zoology* 162: 133–160.

Hamilton, W. D. (1964) The genetical theory of social behaviour I and II. *Journal of Theoretical Biology* 7: 1–52.

(1967) Extraordinary sex ratios. *Science* 156: 477–487.

(1975) Innate social aptitudes of man: an approach from evolutionary genetics, in R. Fox (ed.), *Biosocial anthropology*. London: Malaby Press, 133–155.

Hampshire, S. (1961) *Feeling and expression: inaugural lecture University College London*. Edinburgh: T. & A. Constable Ltd.

(1965) *Freedom of the individual*. London: Chatto & Windus.

Hanfling, O. (2000) *Philosophy and ordinary language*. London: Routledge & Kegan Paul.

(2003) Learning about right and wrong: ethics and language. *Philosophy* 78: 25–41.

Hauser, M. (2006) *Moral minds: how nature designed our universal sense of right and wrong*. New York: Harper Collins.

Hauser, M., Chomsky, N. and Fitch, W. T. (2002) The faculty of language: what is it, who has it, and how did it evolve? *Science* 298: 1565–1576.

Hawkes, K. (1993) Why hunter-gatherers work. *Current Anthropology* 34: 341–361.

Henning, S. J. (1981) Postnatal development: coordination of feeding, digestion, and metabolism. *American Journal of Physiology* 241: G199–G214.

Hill, K. R., Walker, R. S., Božičević, M., Eder, J., Headland, T., Hewlett, B., Hurtado, A. M., Marlowe, F., Wiessner, P. and Wood, B. (2011) Co-residence patterns in hunter-gatherer societies show unique human social structure. *Science* 331: 286–289.

Hirschhorn, R., Yang, D. R., Puck, J. M., Huie, M. L., Jiang, C-K. and Kurlandsky, L. E. (1996) Spontaneous in vivo reversion to normal of an inherited mutation in a patient with adenosine deaminase deficiency. *Nature Genetics* 13: 290–295.

Hoekstra, R. (1990) Evolution of uniparental inheritance of cytoplasmic DNA, in J. Maynard Smith (ed.), *Organizational constraints on the dynamics of evolution*. Manchester University Press, 269–278.

Hoffman, M. L. (2000) *Empathy and moral development*. Cambridge University Press.

Holloway, R. L. (1981) Culture, symbols, and human brain evolution: a synthesis. *Dialectical Anthropology* 5: 287–303.

Hooff, J. A. R. A. M. van (1972) A comparative approach to the phylogeny of laughter and smiling, in R. Hinde (ed.), *Nonverbal communication*. Cambridge University Press, 209–241.

Horsler, K. and Oliver, C. (2006) Environmental influences on the behavioral phenotype of Angelman syndrome. *American Journal of Mental Retardation* 111: 311–321.

Hrdy, S. B. (1999) *Mother nature: natural selection and the female of the species*. London: Chatto & Windus.

(2009) *Mothers and others: the evolutionary origins of mutual understanding*. Cambridge, MA: Harvard University Press.

Humphrey, L. T. (2010) Weaning behaviour in human evolution. *Seminars in Cell and Developmental Biology* 21: 453–461.

Hurford, J. R. (2007) *The origins of meaning: language in the light of evolution 1*. Oxford University Press.

(2012) *The origins of grammar: language in the light of evolution 2*. Oxford University Press.

Hyman, J. (1989) *The imitation of nature*. Oxford: Blackwell.

(1991) Visual experience and blindsight, in J. Hyman (ed.), *Investigating psychology: sciences of the mind after Wittgenstein*. London: Routledge & Kegan Paul, 166–200.

Itier, J-M., Tremp, G. L., Léonard, J-F., Multon, M-C., Ret, G., Schweighoffer, F., Tocqué, B., Bluet-Pajot, M-T., Cormier, V. and Dautry, F. (1998) Imprinted gene in postnatal growth role. *Nature* 393: 125–126.

Jablonka, E. and Lamb, M. J. (1995) *Epigenetic inheritance and evolution*. Oxford University Press.

(2005) *Evolution in four dimensions*. Cambridge, MA: MIT Press.

Jacob, F and Monod, J. (1961). Genetic regulatory mechanisms in the synthesis of proteins. *Journal of Molecular Biology* 3: 318–356.

James, W. (1950 [1890]) *The principles of psychology*, 2 vols. New York: Dover.

Jauniaux, E., Gulbis, B. and Burton, G. J. (2003) The human first trimester gestational sac limits rather than facilitates oxygen transfer to the foetus: a review. *Placenta* 24 (Supplement A): S86–S93.

Jerne, N. K. (1955) The natural-selection theory of antibody formation. *Proceedings of the National Academy of Sciences USA* 41: 849–857.

(1967) Antibodies and learning: selection versus instruction, in G. C. Quarton, T. Melnechuck and F. O. Schmitt (eds.), *The neurosciences: a study program*. New York: Rockerfeller University Press, 200–205.

Johnston, P. (1999) *The contradictions of modern moral philosophy*. London: Routledge & Kegan Paul.

Joleff, N. and Ryan, M. (1993) Communication development in Angelman syndrome. *Archives of Disease in Childhood* 69: 148–150.

Keller, L. (ed.) (1999) *Levels of selection in evolution*. Princeton University Press.

Kelsey, G. and Feil, R. (2013) New insights into establishment and maintenance of DNA methylation imprints in mammals. *Philosophical Transactions of the Royal Society B* 368, doi: 10.1098/rstb.2011.0336.

Kennedy, G. E. (2005). From the ape's dilemma to the weanling's dilemma: early weaning and its evolutionary context. *Journal of Human Evolution* 48: 123–145.

Kenny, A. (1963) *Action, emotion, and the will*. London: Routledge & Kegan Paul.

(1968) *Descartes*. New York: Random House.

(1975) *Will, freedom and power*. Oxford: Blackwell.

(1984). The homunculus fallacy, in *The legacy of Wittgenstein*. Oxford: Blackwell, 125–136. First published in M. Greene (ed.), *Interpretations of life and mind*. London: Routledge & Kegan Paul, 1971.

(1988a) Cosmological explanation and understanding, in L. Herzberg and J. Pietarinen (eds.), *Perspectives on human conduct*. Leiden: Brill, 72–87.

(1988b) *The self: Aquinas lecture 1988*. Milwaukee, WI: Marquette University Press.

(1989) *The metaphysics of mind*. London: Clarendon Press.

Keverne, E. B., Fundele, R., Narasimha, M., Barton, S. C. and Surani, M. A. (1996) Genomic imprinting and the differential roles of parental genomes in brain development. *Developmental Brain Research* 92: 91–100.

Kimura, M. (1983) *The neutral theory of molecular evolution*. Cambridge University Press.

King, A. P. and West, M. J. (1977) Species identification in the Northern American cowbird: appropriate responses to abnormal song. *Science* 195: 1002–1004.

Kirkwood, T. B. L. and Austad, S. N. (2000) Why do we age? *Nature* 408: 233–238.

Kirschner, M. and Gerhart, J. C. (2005) *The plausibility of life: resolving Darwin's dilemma*. New Haven, CT: Yale University Press.

Kozlov, S. V., Bogenpohl, J. W., Howell, M. P., Wevrick, R., Panda, S., Hogenesch, J. B., Muglia, L. J., van Gelder, R. N., Herzog, E. D. and Stewart, C. L. (2007) The imprinted gene Magel2 regulates normal circadian output. *Nature Genetics* 39: 1266–1272.

Kühnle, S., Mothes, B., Matentzoglu, K. and Scheffner, M. (2013) Role of the ubiquitin ligase E6AP/UBE3A in controlling levels of the synaptic protein Arc. *Proceedings of the National Academy of Sciences USA*. In press.

Lachmann, M., Számadó, S., and Bergstrom, C. T. (2001) Cost and conflict in animal signals and human language. *Proceedings of the National Academy of Sciences USA* 98: 13189–13194.

Leavitt, G. C. (1990) Sociobiological explanations of incest avoidance: a critical review of evidential claims. *American Amthropologist* 92: 971–993.

(2007) The incest taboo? A reconsideration of Westermarck. *Anthropological Theory* 7: 393–419.

Lederberg, J. (2002) Instructive selection and immunological theory. *Immunological Reviews* 185: 50–53.

Lehrman, D. S. (1953) A critique of Konrad Lorenz' theory of instinctive behaviour. *Quarterly Review of Biology* 28: 337–363.

(1970) Semantic and conceptual issues in the nature–nurture problem, in L. R. Aronson, E. Tobach, D. S. Lehrman and J. S. Rosenblatt (eds.), *Development and the evolution of behaviour: essays in memory of T. C. Schneirla*. San Francisco: W. H. Freeman, 17–52.

Leigh, E. G. (1977) How does selection reconcile individual advantage with the good of the group? *Proceedings of the National Academy of Sciences USA* 74: 4542–4546.

Lennox, J. G. (1992) Teleology, in E. Fox Keller and E. A. Lloyd (eds.), *Keywords in evolutionary biology*. Cambridge, MA: Harvard University Press, 324–333.

Lévi-Strauss, C. (1969 [1949]) *The elementary structures of kinship*. Boston, MA: Beacon Press. Translation of *Les structures élémentaires de la parenté* by James Harle Bell, John Richard von Sturmer and Rodney Needham. Paris: Presses Universitaires de France.

Lieberman, D., Tooby, J. and Cosmides, L. (2007) The architecture of human kin recognition. *Nature* 445: 727–731.

Locke, J. L. (2006) Parental selection of vocal behavior: crying, cooing, babbling, and the evolution of language. *Human Nature* 17: 155–168.

Lorenz, K. (1965) *Evolution and modification of behaviour*. University of Chicago Press.

   (1977) *Behind the mirror: a search for a natural history of human knowledge*. London: Methuen & Co. Ltd.

Luria, S. E. and Delbrück, M. (1943) Mutations of bacteria from virus sensitivity to virus resistance. *Genetics* 28: 491–511.

Lyons, D. E., Young, A. G. and Keil, F. C. (2007) The hidden structure of overimitation. *Proceedings of the National Academy of Sciences USA* 104: 19751–19756.

Malcolm, N. (1982) The relation of language to instinctive behaviour. *Philosophical Investigations* 5: 3–22.

Manda, S. O. M. (1999) Birth intervals, breastfeeding and determinants of childhood mortality in Malawi. *Social Science and Medicine* 58: 301–312.

Manning, A. (1972 [1967]) *An introduction to animal behaviour*, second edition. London: Edward Arnold.

Martin, G. B. and Clark III, R. D., (1982) Distress crying in neonates: species and peer specificity. *Developmental Psychology* 18: 3–9.

Marx, K. (1867) *Das Kapital: Kritik der politischen Oekonomie*. Hamburg: Verlag von Otto Meissner.

Maynard Smith, J. (1986). *The problems of biology*. Oxford University Press.

Maynard Smith, J. and Szathmáry, E. (1995) *The major transitions in evolution*. New York: W. H. Freeman.

   (1999) *The origins of life: from the birth of life to the origin of language*. Oxford University Press.

Mayr, E. (1961) Cause and effect in biology. *Science* 134: 1501–1506.

   (1992) The idea of teleology. *Journal of the History of Ideas* 35: 117–135.

McCallum, H. and Jones, M. (2006) To lose both would look like carelessness: Tasmanian devil facial tumour disease, *PLoS Biology* 4: 1671–1674.

Megone, C. (1998) Aristotle's function argument and the concept of mental health. *Philosophy, Psychiatry & Psychology* 5: 187–201.

   (2000) Mental illness, human function and values. *Philosophy, Psychiatry & Psychology* 7: 45–65.

Meyer-Lindenberg, A., Buckholtz, J. W., Kolachana, B. R., Hariri, A. R., Pezawas, L., Blasi, G., Wabnitz, A., Honea, R., Verchinski, B., Callicott, J. H., Egan, M., Mattay, V. and Weinberger, D. R. (2006)

Neural mechanisms of genetic risk for impulsivity and violence in humans. *Proceedings of the National Academy of Sciences USA* 103: 6269–6274.

Michod R. E., Viossat, Y., Solari, C. A., Hurand, M. and Nedelcu, A. M. (2006) Life history evolution and the evolution of multicellularity. *Journal of Theoretical Biology* 239: 257–272.

Mikhail, J. (2007) Universal moral grammar: theory, evidence and the future. *Trends in Cognitive Sciences* 11: 143–152.

Mills, W. and Moore, T. (2004) Polyandry, life-history trade-offs, and the evolution of imprinting at Mendelian loci. *Genetics* 168: 2317–1327.

Mochizuki, A., Takeda, Y. and Iwasa, Y. (1996) The evolution of genomic imprinting. *Genetics* 144: 1283–1295.

Neff, B. D. (2003) Decisions about parental care in response to perceived paternity. *Nature* 422: 716–719.

Nesse, R. M. and Williams, G. C. (1994) *Why we get sick: the new science of Darwinian medicine*. New York: Times Books.

Nowak, M. A. and Sigmund, K. (2005) Evolution of indirect reciprocity. *Nature* 437: 1291–1298.

Ohlsson, R. (ed.) (1999) *Genomic imprinting: an interdisciplinary approach*. Berlin: Springer-Verlag.

Ohlsson, R., Hall, K. and Ritzen, M. (eds.) (1995) *Genomic imprinting: causes and consequences*. Cambridge University Press.

Okasha, S. (2006) *Evolution and the levels of selection*. Oxford University Press.
  (2008) Fisher's fundamental theorem of natural selection: a philosophical analysis. *British Journal for the Philosophy of Science* 59: 319–351.

Olson, M. (1965) *The logic of collective action*. Cambridge, MA: Harvard University Press.

Paley, W. (2006 [1802]) *Natural theology, or evidence of the existence and attributes of the Deity collected from the appearances of nature*, edited by M. D. Eddy and D. Knight. Oxford University Press.

Pauling, L. (1940) A theory of the structure and process of formation of antibodies. *Journal of the American Chemical Society* 62: 2643–2657.

Pena de Ortiz, S. and Arshavsky, Y. I. (2001) DNA recombination as a possible mechanism in declarative memory: a hypothesis. *Journal of Neuroscience Research* 63: 72–81.

Pinker, S. (1994) *The language instinct*. New York: W. Morrow and Company Inc.
  (2002) *The blank slate: the modern denial of human nature*. London: Penguin Books.

Plagge, A., Gordon, E., Dean, W., Boiani, R., Cinti, S., Peters, J. and Kelsey, G. (2004) The imprinted signaling protein XLαs is required for postnatal adaptation to feeding. *Nature Genetics* 36: 818–826.

Plagge, A., Isles, A. R., Gordon, E., Humby, T., Dean, W., Gritsch, S., Fischer-Colbrie, R., Wilkinson, L. S. and Kelsey, G. (2005) Imprinted NESP55 influences behavioral reactivity to novel environments. *Molecular and Cellular Biology* 25: 3019–3026.

Price, G. R. (1972) Fisher's 'fundamental theorem' made clear. *Annals of Human Genetics* 36: 129–140.

Prinz, J. (2007) *The emotional construction of morals*. Oxford University Press.

Profet, M. (1992) Pregnancy sickness as adaptation: a deterrent to maternal ingestion of teratogens, in J. Barkow, L. Cosmides and J. Tooby (eds.), *The adapted mind*. Oxford University Press, 327–365.

Queller, D. C. (1997) Cooperators since life began. *Quarterly Review of Biology* 72: 184–188.

(2000) Relatedness and the fraternal major transitions. *Philosophical Transactions of the Royal Society B* 355: 1647–1655.

(2003) Theory of genomic imprinting conflict in social insects. *BMC Evolutionary Biology* 3: 15.

Queller, D. C. and Goodnight, K. F. (1989) Estimating relatedness using genetic markers. *Evolution* 43: 258–275.

Queller, D. C., Ponte, E., Bozzaro, S. and Strassmann, J. E. (2003) Single-gene greenbeard effects in the social amoeba *Dictyostelium discoideum*. *Science* 299: 105–106.

Queller, D. C. and Strassmann, J. E. (2009) Beyond society: the evolution of organismality. *Philosophical Transactions of the Royal Society B* 364: 3143–3155.

Rapkins, R. W., Hore, T., Smithwick, M., Ager, B., Pask, A. J., Renfee, M. B., Kohn, M., Hameister, H., Nicholls, R. D., Deakin, J. E. and Marshall Graves, J. A. (2006) Recent assembly of an imprinted domain from non-imprinted components. *PloS Genetics* 2: 1666–1675.

Ratnieks, F. L. W., Wenseleers, T. and Foster, K. R. (2006) Conflict resolution in insect societies. *Annual Review of Entomology* 51: 581–608.

Reik, W. and Surani, A. (eds.) (1997) *Genomic imprinting*. Oxford University Press.

Reik, W. and Walter, J. (2001) Genomic imprinting: parental influence on the genome. *Nature Reviews Genetics* 2: 21–32.

Rice, W. R. (1984) Sex chromosomes and the evolution of sexual dimorphism. *Evolution* 38: 735–742.

Ridley, M. (2000) *Mendel's demon: gene justice and the complexity of life*. London: Weidenfeld & Nicolson.

Ridley, M. and Grafen, A. (1981) Are green beard genes outlaws? *Animal Behavior* 29: 954–955.

Rundle, B. (1997) *Mind in action*. Oxford: Clarendon Press.

Ruse, M. (2003) *Darwin and design: does evolution have a purpose?* Cambridge, MA: Harvard University Press.

Russell, B. (1967 [1912]) *The problems of philosophy*. Oxford University Press.

Ryle, G. (1980 [1949]) *The concept of mind*. Harmondsworth: Penguin Books.

(2009 [1971]) *Collected essays: 1929–1968*. Abingdon: Routledge & Kegan Paul. Reprint of the first edition with a foreword by Julia Tanney.

Sagi, A. and Hoffman, M. L. (1976) Empathic distress in the newborn. *Developmental Psychology* 12: 175–176.

Schaller, F., Watrin, F., Sturny, R., Massacrier, A., Szepetowski, P. and Muscatelli, F. (2010) A single postnatal injection of oxytocin rescues the lethal feeding behaviour in mouse newborns deficient for the imprinted Magel2 gene. *Human Molecular Genetics* 19: 4895–4905.

Schleidt, W., Schleidt, M. and Magg, M (1960) Störung der Mutter-Kind-Beziehung bei Truthühnern durch Gehöreverlust. *Behaviour* 16: 254–260.

Schroeder, S. (2006) *Wittgenstein: the way out of the fly bottle.* Cambridge: Polity Press.

Scott-Phillips, T. C. (2008) On the correct application of animal signalling theory to human communication, in A. D. M. Smith, K. Smith and R. Ferrer i Cancho (eds.), *The evolution of language: proceedings of the 7th international conference (EVOLANG7).* Singapore: World Scientific, 275–282.

(2010) Evolutionary psychology and the origins of language. *Journal of Evolutionary Psychology* 8: 289–307.

Sellen, D. W. (2007) Evolution of infant and young child feeding: Implications for contemporary public health. *Annual Review of Nutrition* 27: 123–148.

Shahidullah, M. (1994) Breast-feeding and child survival in Matlab, Bangladesh. *Journal of Biosocial Science* 26: 143–154.

Shavit, Y., Fischer, C. S. and Koresh, Y. (1994) Kin and nonkin under collective threat: Israeli networks during the Gulf War. *Social Forces* 72: 197–215.

Shor, E. and Simchai, D. (2009) Incest avoidance, the incest taboo, and social cohesion: revisiting Westermarck and the case of the Israel kibbutzim. *American Journal of Sociology* 114: 1803–1842.

Siemensma, E. P., de Lind van Wijngaarden, R. F., Otten, B. J., de Jong, F. H. and Hokken-Koelega, A. C. (2011) Pubarche and serum dehydroepiandrosterone sulphate levels in children with Prader–Willi syndrome. *Clinical Endocrinology* 75: 83–89.

Simner, M. L. (1971) Newborn's response to the cry of another infant. *Developmental Psychology* 5: 136–150.

Skyrms, B. (2004) *The stag hunt and the evolution of social structure.* Cambridge University Press.

Smit, H. (1989) *De biologie en methodologie van aanleg en omgeving.* Groningen: Wolters-Noordhoff.

(1995a) Are animal displays bodily movements or manifestations of the animals' mind? *Behavior and Philosophy* 23: 15–21.

(1995b) Zwangerschapsmisselijkheid in een evolutionair perspectief. *De Psycholoog* 11: 449–455.

(2002) De seksespecifieke erfenis van de ouders. *Nederlands Tijdschrift voor de Psychologie* 57: 82–94.

(2005) De evolutionaire genetica van psychopathologie. *Nederlands Tijdschrift voor de Psychologie* 60: 27–43.

(2006) Tirannieke mechanismen in het brein: Prader–Willi-syndrom en Angelman-syndroom. *Nederlands Tijdschrift voor de Psychologie* 61: 41–53.

(2007) Conflicten in het brein. *Algemeen Nederlands Tijdschrift voor de Wijsbegeerte* 99: 173–187.

(2009) Genomic imprinting and communicative behaviour: Prader–Willi and Angelman syndrome. *Netherlands Journal of Psychology* 65: 78–88.

(2010a) A conceptual contribution to battles in the brain. *Biology and Philosophy* 25: 803–821.

(2010b) Darwin's rehabilitation of teleology versus Williams' replacement of teleology by natural selection. *Biological Theory* 5: 357–365.

(2010c) *Darwinisme, monisme en ziekte*. Amsterdam: Boom.

(2010d) The development of altruistic behaviour out of reactive crying. *Biological Theory* 5: 79–86.

(2012) Eindeloze discussies over het brein. *De Psycholoog* 47: 26–32.

Somel, M., Liu, X. and Khaitovich, P. (2013) Human brain evolution: transcripts, metabolites and their regulators. *Nature Reviews Neuroscience* 14: 112–127.

Spalding, D. (1954 [1873]). Instinct: with original observations on young animals. *British Journal of Animal Behaviour* 2: 2–11. First published in *Macmillan's Magazine* 27: 282–293.

Stanford, C. B. (2003) *Upright: the evolutionary key to becoming human*. Boston, MA: Houghton Mifflin.

Stearns, S. C. and Hoekstra, R. F. (2005), *Evolution: an introduction*, second edition. Oxford University Press.

Stout, D. and Chaminade, T. (2012) Stone tools, language and the brain in human evolution. *Philosophical Transactions of the Royal Society B* 367: 75–87.

Strassmann, J. E. and Queller, D. C. (2010) The social organism: congresses, parties, and committees. *Evolution* 64: 605–616.

Strawson, P. F. (1966) *The bounds of sense*. London: Methuen.

(2008 [1974]) *Freedom and resentment and other essays*. London: Routledge & Kegan Paul.

Sullivan, M. (1986) In what sense is contemporary medicine dualistic? *Culture, Medicine and Psychiatry* 10: 331–350.

Svetlova, M., Nichols, S. and Brownell, C. A. (2010) Toddlers' prosocial behaviour: from instrumental to empathic to altruistic helping. *Child Development* 81: 1814–1827.

Szamado, S., and Szathmary, E. (2006) Selective scenarios for the emergence of natural language. *Trends in Ecology and Evolution* 21: 555–561.

Taylor, D. R., Zeyl, C. and Cooke, E. (2002) Conflicting levels of selection in the accumulation of mitochondrial defects in Saccharomycetes cerevisiae. *Proceedings of the National Academy of Sciences USA* 99: 3690–3694.

Tennie, C., Call, J. and Tomasello, M. (2009) Ratcheting up the ratchet: on the evolution of cumulative culture. *Philosophical Transactions of the Royal Society B* 364: 2405–2415.

Tinbergen, N. (1963) On aims and methods of ethology. *Zeitschrift für Tierpsychologie*. 20: 410–433.

(1989 [1951]) *The study of instinct*. Oxford: Clarendon Press. Reprint of the second edition of 1969.

Ting, J. T. and Feng, G. (2011) Neurobiology of obsessive–compulsive disorder: insights into neural circuitry dysfunction through mouse genetics. *Current Opinion in Neurobiology* 21: 842–848.

Tomasello, M. (2008) *Origins of human communication*. Cambridge, MA: MIT Press.

Tomasello, M. and Call, J. (1997) *Primate cognition*. New York: Oxford University Press.

Tomasello, M., Carpenter, M., Gall, J., Behne, T. and Moll, H. (2005) Understanding and sharing intentions: the origins of cultural cognition. *Behavioral and Brain Sciences* 28: 675–735.

Tooby, J. and Cosmides, L. (1992) The psychological foundations of culture, in J. H. Barkow, L. Cosmides and J. Tooby (eds.), *The adapted mind*. Oxford University Press, 19–36.

(2005) Conceptual foundations of evolutionary psychology, in D. M. Buss (ed.), *A handbook of human evolutionary psychology*. New York: Wiley, 5–67.

Topál, J., Gergely, G., Miklósi, A., Erdőhegyi, Á. and Csibra, G. (2008) Infants' perseverative search errors are induced by pragmatic misinterpretation. *Science* 321: 1831–1834.

Trivers, R. L. (1971) The evolution of social reciprocal altruism. *Quarterly Review of Biology* 46: 35–57.

(1972) Parental investment and sexual selection, in B. Campbell (ed.), *Sexual selection and the descent of man, 1871–1971*. Chicago, IL: Aldine-Atherton, 136–179.

(1974) Parent–offspring conflict. *American Zoologist* 14: 249–264.

(1985) *Social evolution*. Menlo Park, CA: The Benjamin-Cummings Publishing Company.

(1997) Genetic basis of intrapsychic conflict, in N. L. Segal, G. E. Weisfeld and C. C. Weisfeld (eds.), *Uniting psychology and biology: integrative perspectives on human development*. Washington, DC: American Psychological Association, 385–395.

(2000) The elements of a theory of self-deception, *Annals of the New York Academy of Science* 907: 114–131.

Tseng, Y-H., Butte, A. J., Kokkotou, E., Yechoor, V. K., Taniguchi, C. M., Kriauciunas, K. M., Cypess, A. M., Niinobe, M. N., Yoshikawa, K., Patti, M. E. and Kahn, C. R. (2005) Prediction of preadipocyte differentiation by gene expression eveals role of insulin receptor substrates and necdin. *Nature Cell Biology* 7: 601–611.

Úbeda, F. (2008) Evolution of genomic imprinting with biparental care: implications for Prader–Willi and Angelman syndromes. *PLoS Biology* 6: 1678–1692.

Úbeda, F. and Gardner, A. (2010) A model for genomic imprinting in the social brain: juveniles. *Evolution* 64: 2587–2600.

(2011) A model for genomic imprinting in the social brain: adults. *Evolution* 65: 462–475.

Viding, E., Jones, A. P., Frick, P. J., Moffit, T. E. and Plomin, R. (2008) Heritability of antisocial behaviour at 9: do callous-unemotional traits matter? *Developmental Science* 11: 17–22.

Visser, M. E, Lessells, C. M. (2001) The costs of egg production and incubation in great tits (*Parus major*). *Proceedings of the Royal Society B* 268: 1271–1277.

Warneken, F. and Tomasello, M. (2009) Varieties of altruism in children and chimpanzees. *Trends in Cognitive Science* 23: 397–402.

Weismann, A. (1983 [1904]) *The theory of evolution*, 2 vols. New York: AMS Press. Translated by J. A. Thomson and M. R. Thomson from the 1904 edition, *Vorträge über Descendenztheorie*.

West, S. A., El Mouden, C. and Gardner, A. (2011) Sixteen common misconceptions about the evolution of cooperation in humans. *Evolution and Human Behaviour* 32: 231–262.

West, S. A., Griffin, A. S. and Gardner, A. (2007) Evolutionary explanations for cooperation, *Current Biology* 17: R661–R672.

Westermarck, E. (1906) *The origin and development of moral ideas*, vol. 1. London: Macmillan.

(1925 [1891]) *The history of human marriage*, fifth, rewritten edition, vol. 2. London: Macmillan.

White, A. R. (1982). *The nature of knowledge*. Totowa, NJ: Rowman and Littlefield.

Wiggins, D. (2001) *Sameness and substance renewed*. Cambridge University Press.

Wilkins, J. F. (ed.) (2009) *Genomic imprinting: advances in experimental medicine and biology*, vol. 626. New York: Springer.

Wilkins, J. F. and Haig, D. (2003) What good is genomic imprinting: the function of parent-specific gene expression. *Nature Reviews Genetics* 4: 1–10.

Wilkinson, L. S., Davies, W., and Isles, A. R. (2007) Genomic imprinting effects of brain development and function. *Nature Reviews Neuroscience* 8: 832–843.

Williams, C. A., Beaudet, A. L., Clayton-Smith, J., Knoll, J. H., Kyllerman, M., Laan, L. A., Magenis, R. E., Moncla, A., Schinzel, A. A., Summers, J. A. and Wagstaff, J. (2006) Angelman syndrome 2005: updated consensus for diagnostic criteria. *American Journal of Medical Genetics* 140A: 413–418.

Williams, G. C. (1985) A defense of reductionism in evolutionary biology, in R. Dawkins and M. Ridley (eds.), *Oxford surveys in evolutionary biology*. Oxford University Press, 1–27.

(1992) *Natural selection: domains, levels, and challenges*. Oxford University Press.

(1996 [1966]) *Adaptation and natural selection*, second edition. Princeton University Press.

Williams, L. M. and Finkelhor, D. (1995) Paternal caregiving and incest: test of a biosocial model. *American Journal of Orthopsychiatry* 65: 101–113.

Wittgenstein, L. (1958) *The blue and brown books*. Oxford: Blackwell.

(2009 [1953]) *Philosophical investigations*, translated by G. E. M. Anscombe, P. M. S. Hacker and J. Schulte. Revised fourth edition by P. M. S. Hacker and J. Schulte. Chichester: Wiley-Blackwell.

Wolf, A. P. and Durham, W. D. (eds.) (2005) *Inbreeding, incest and the incest taboo.* Stanford University Press.

Wright, G. H. von (1963) *Varieties of goodness.* London: Routledge & Kegan Paul.

(1974) *Causality and determinism.* New York: Columbia University Press.

Wynne-Edwards, V. C. (1962) *Animal dispersion in relation to social behavior.* London: Oliver & Boyd.

Yashiro, K., Riday, T., Condon, K., Roberts, A. C., Bernardo, D. R., Prakash, R., Weinberg, R. J., Ehlers, M. D. and Philpot, B. D. (2009) Ube3A is required for experience-dependent maturation of the neocortex. *Nature Reviews Neuroscience* 12: 777–783.

Yu, S., Gavrilova, O., Chen, H., Lee, R., Liu, J., Pacak, K., Parlow, A. F., Quon, M. J., Reitman, M. L. and Weinstein, L. S. (2000) Paternal versus maternal transmission of a stimulatory G-protein α subunit knockout produces opposite effects on energy metabolism. *Journal of Clinical Investigation* 105: 615–623.

Zahn-Waxler, C., Radke-Yarrow, M., Wagner, E. (1992) Development of concern for others. *Developmental Psychology* 28: 126–136.

Zajonc, R. B. (1984) On the primacy of affect. *American Psychologist* 39: 117–124.

Index

Printed in the United States
By Bookmasters